The Cambridge Companion to
John Herschel

It has been said that being scientific in Victorian England meant to be as much like John Herschel as possible. This volume shows readers what it meant to be John Herschel (1792–1871), one of England's most prominent polymaths. Drawing on his published oeuvre and recent scholarship, as well as an immense amount of surviving archival material and correspondence, these essays present the first ever comprehensive account of Herschel's life, work, and legacy. From mathematics and astronomy, to philosophy and politics, the volume sheds new light on his crucial role in the history of Victorian science and explores a wide array of issues in the history of nineteenth-century culture, philosophy, mathematics, and beyond.

STEPHEN CASE is a professor in the Department of Chemistry and the Geosciences, Olivet Nazarene University, and the author of *Making Stars Physical: The Astronomy of Sir John Herschel* (2018).

LUKAS M. VERBURGT is Gerda Henkel Stiftung Research Scholar and affiliated to the Netherlands Institute for Advanced Study. He has authored and coedited several books and numerous articles on the history of science and philosophy in nineteenth-century Britain.

The Cambridge Companion to
John Herschel

Edited by

STEPHEN CASE
Olivet Nazarene University

LUKAS M. VERBURGT
Netherlands Institute for Advanced Study in the Humanities
and Social Sciences

CAMBRIDGE
UNIVERSITY PRESS

Shaftesbury Road, Cambridge CB2 8EA, United Kingdom

One Liberty Plaza, 20th Floor, New York, NY 10006, USA

477 Williamstown Road, Port Melbourne, VIC 3207, Australia

314–321, 3rd Floor, Plot 3, Splendor Forum, Jasola District Centre, New Delhi – 110025, India

103 Penang Road, #05-06/07, Visioncrest Commercial, Singapore 238467

Cambridge University Press is part of Cambridge University Press & Assessment,
a department of the University of Cambridge.

We share the University's mission to contribute to society through the pursuit of education,
learning and research at the highest international levels of excellence.

www.cambridge.org
Information on this title: www.cambridge.org/9781009237703

DOI: 10.1017/9781009237727

First published 2024

A catalogue record for this publication is available from the British Library

Library of Congress Cataloging-in-Publication Data
NAMES: Case, Stephen, editor. | Verburgt, Lukas M., editor.
TITLE: The Cambridge companion to John Herschel / edited by Stephen Case, Olivet Nazarene
University, Lukas Verburgt, Netherlands Institute for Advanced Study.
DESCRIPTION: Cambridge, United Kingdom ; New York, NY : Cambridge University Press,
2024. | Includes bibliographical references and index.
IDENTIFIERS: LCCN 2023034830 (print) | LCCN 2023034831 (ebook) | ISBN 9781009237703
(hardcover) | ISBN 9781009237673 (paperback) | ISBN 9781009237727 (ebook)
SUBJECTS: LCSH: Herschel, John F. W. (John Frederick William), 1792-1871. | Astronomers-
Great Britain–Biography | Astronomy–History–19th century. | Astronomers–South Africa-
Biography. | Astronomy–South Africa–History–19th century.
CLASSIFICATION: LCC QB36.H59 C36 2024 (print) | LCC QB36.H59 (ebook) |
DDC 520.92 [B]–dc23
LC record available at https://lccn.loc.gov/2023034830
LC ebook record available at https://lccn.loc.gov/2023034831

ISBN 978-1-009-23770-3 Hardback
ISBN 978-1-009-23767-3 Paperback

Contents

Figures

Contributors

WILLIAM J. ASHWORTH is Reader in History at the University of Liverpool. He is the author of *Customs and Excise: Trade, Production and Consumption in England 1640–1845* (2003), *The Industrial Revolution: The State, Knowledge and Global Trade* (2017), and *The Trinity Circle: Anxiety, Intelligence, and Knowledge Creation in Nineteenth-Century England* (2021).

STEPHEN CASE is a professor in the Department of Chemistry and the Geosciences at Olivet Nazarene University. He is the author of *Making Stars Physical: The Astronomy of Sir John Herschel* (2018) and the forthcoming *Creatures of Reason: John Herschel and the Invention of Science*.

TONY CRILLY is Emeritus Reader in Mathematical Sciences at Middlesex University. His principal research interest is the history of mathematics. He has written and edited works on fractals, chaos, and computing, is the author of *Arthur Cayley: Mathematician Laureate of the Victorian Era* (2005), and is completing a biography of mathematician Thomas P. Kirkman.

EDWARD J. GILLIN is Lecturer in the History of Building Sciences and Technology at the Bartlett School of Sustainable Construction, UCL. His books include *Entente Imperial: British and French Power in the Age of Empire* (2022), *Sound Authorities: Scientific and Musical Knowledge in Nineteenth-Century Britain* (2022), and *An Empire of Magnetism: Global Science and the British Magnetic Enterprise in the Age of Imperialism* (2024).

GREGORY A. GOOD is Director Emeritus of the Center for History of Physics, American Institute of Physics. He writes about the history of the earth sciences from the time of John Herschel to the twenty-first century. His current books are on astronomers' research on the earth and on space weather.

CAROLIN LANGE is an artist specializing in experimental photography and spectroscopy of the 19th century. Her work has been exhibited internationally at the Platform for Contemporary Art TENT and at the Center for Art and Media ZKM, among others. She is currently an M4C funded PhD candidate at the Photographic History Research Centre at De Montfort University, Leicester.

OMAR W. NASIM is Professor of the History of Science at the University of Regensburg in Germany. Nasim's research centers on the formation of knowledge historically considered, especially at the intersections of material, visual, and intellectual cultures. He is the author of the award-winning book *Observing by Hand: Sketching the Nebulae in the Nineteenth Century* (2013) and, more recently, *The Astronomer's Chair: A Visual and Cultural History* (2021).

CHARLES H. PENCE is a lecturer at the Université catholique de Louvain in Louvain-la-Neuve, Belgium, where he directs the Center for Philosophy of Science and Societies (CEFISES). He also serves as a coeditor of the journal *Philosophy, Theory, and Practice in Biology* (*PTPBio*). His work centers on the philosophy and history of biology, with a focus on the introduction and contemporary use of chance and statistics in evolutionary theory.

STEVEN RUSKIN'S work focuses on the history of nineteenth-century science, specifically astronomy, exploration, and the field sciences. He is the author of *John Herschel's Cape Voyage: Private Science, Public Imagination, and the Ambitions of Empire* (2004) and *America's First Great Eclipse: How Scientists, Tourists, and the Rocky Mountain Eclipse of 1878 Changed Astronomy Forever* (2017).

LUKAS M. VERBURGT is currently a Gerda Henkel Stiftung Research Scholar and affiliated as Research and Project Associate to the Netherlands Institute for Advanced Study in the Humanities and Social Sciences. His main research focus is the changing relationship between science and philosophy in the long nineteenth century, especially in Victorian Britain. He is the author of numerous articles and (co)editor of several volumes, including *Aristotle's Syllogism and the Creation of Modern Logic* (2023) and the forthcoming *Cambridge Companion to Charles Babbage*.

KELLEY WILDER is Professor of Photographic History and Director of the Photographic History Research Centre at De Montfort University, Leicester, UK. She has published widely on material cultures of nineteenth- and twentieth-century photography and science, including *Photography and Science* (2009) and *Documenting the World: Film, Photography and the Scientific Record* (2016, with Gregg Mitman).

EMILY WINTERBURN has served as Curator at the Royal Observatory, Greenwich, and as a research fellow at the Centre for History and Philosophy of Science, University of Leeds. She is the author of *The Stargazer's Guide: How to Read Our Night Sky* (2009) and *The Quiet Revolution of Caroline Herschel: The Lost Heroine of Astronomy* (2017).

Acknowledgments

We would like to express our gratitude to the late Michael Hoskin and to Michael Crowe, who together have done much to move Herschel scholarship forward since the early work of Susan Faye Canon. We acknowledge permission from the following institutions to quote from materials held in their collections: the British Library; the Harry Ransom Center, University of Texas, Austin; National Maritime Museum; Royal Society; Science Museum Group; St John's College and Trinity College, Cambridge; and Yale University. Lukas Verburgt acknowledges support from a Gerda Henkel Stiftung research grant (AZ 23/F22).

Chronology

1792   Born, March 7, only child of William Herschel and Mary Pitt (née Baldwin) in Slough

1800   Attended Eton College for three months

1800–09   Educated at a private school in Hitcham, near Slough, led by George Gretton, graduate of Trinity College, Cambridge, and later dean of Hereford

1802   Traveled with his parents to Paris, where his father and others, including Pierre-Simon Laplace, had an audience with Napoleon Bonaparte

1809   Entered St John's College, Cambridge

1809–10   Occasional factory visits around Britain with his father, recorded in a "Travel Diary"

1813   Graduated Senior Wrangler; awarded First Smith's Prize; elected fellow of St. John's College (relinquished upon his marriage in 1829); elected fellow of the Royal Society; founding member and first president of the Analytical Society; publication of *Memoirs of the Analytical Society*, composed together with Charles Babbage

1814   Entered Lincoln's Inn to train for the Bar

1815   Returned to Cambridge as mathematics "subtutor" and examiner at St. John's College

1816   Publication of translation of S. F. Lacroix, *An Elementary Treatise on the Differential and Integral Calculus*, cotranslated with Babbage and George Peacock; left Cambridge in October for the family house in Slough

("Observatory House") to be "going, under his father's
direction, to take up star-gazing"

1820   Publication of *A Collection of Examples of the Application of
the Calculus of Finite Differences*; founding member of the
Astronomical Society of London, renamed Royal
Astronomical Society in 1831

1821   Awarded Royal Society's Copley Medal for mathematical
contributions to its transactions

1821–23   Collaboration with James South on the (re)observation of
double stars

1822   Death of father, William Herschel, aged 83

1823   Publication of work on optical spectra of metal salts in
Transactions of the Royal Society of Edinburgh

1824–27   Secretary of the Royal Society

1825   Awarded French Academy of Sciences' Lalande Prize with
James South; invented the actinometer, to measure heating
power of the sun's rays

1826   Awarded Astronomical Society's gold medal for work with
his father

1827   Publication of "Light" in the *Encyclopaedia Metroplitana*

1827–29   First period as president of the Astronomical Society, later
periods followed in 1839–41 and 1847–49

1829   Marriage to Margaret Brodie Stewart (1810–84)

1830   Candidate for presidency of the Royal Society, lost to the
duke of Sussex; birth of first of a total of twelve children,
Caroline Herschel (†1909)

1831   Publication of *A Preliminary Discourse on the Study of
Natural Philosophy*; made Knight of the Royal
Guelphic Order

1832   Elected Foreign Honorary Member of the American
Academy of Arts and Sciences

1833   Publication of *A Treatise on Astronomy*; awarded French
Academy of Sciences' Lalande Prize

1834–38   Lived and worked at Feldhausen, Cape Town, South Africa,
where he erected a twenty-foot reflecting telescope, the site
of which was marked by a stone obelisk in 1842; produced
131 botanical illustrations with his wife, using a camera
lucida (later published as *Flora Herscheliana*)

1836   Awarded Astronomical Society's gold medal; elected Foreign Honorary Member of the Royal Swedish Academy of Sciences; birth of second son, Alexander Stewart Herschel (†1907), astronomer

1837   Endorsement of Charles Lyell's gradualist (or "uniformitarian") view of geological change, included in Babbage's unofficial *Ninth Bridgewater Treatise*

1838   Created baronet of Slough in the county of Buckingham

1839   First glass-plate photograph, showing his father's forty-foot telescope

1839–40   Read two papers on "photography" (a term of his coinage) to the Royal Society

1840   Awarded Royal Society's gold medal for contributions to the development of photographic techniques

1841   Birth of fifth daughter, Amelia Herschel (†1926), with William Whewell as godfather

1842   Invented the cyanotype process (blueprints)

1847   Publication of *Results of Astronomical Observations made at the Cape of Good Hope*; awarded Royal Society's Copley Medal for Cape "Results"

1849   Publication of *Outlines of Astronomy*

1850–54   Master of the Mint

1854   Elected member of the American Philosophical Society

1855   Birth of ninth daughter, Constance Anne Herschel (†1939), later lecturer in natural sciences at Girton College, Cambridge

1859   Published a study of color blindness in *Philosophical Magazine*

1864   Publication of *General Catalogue of Nebulae and Clusters*

1865–71   Compiled a catalog of all known double and multiple star systems, which appeared posthumously in 1874 as *General Catalogue of 10,300 Multiple and Double Stars*

1866   Publication of *The Iliad of Homer: Translated into English Accentuated Hexameters*

1871   Died, May 11, at home in Collingwood, Kent, aged 79; received a national funeral and buried in Westminster Abbey, London

STEPHEN CASE AND LUKAS M. VERBURGT

Introduction

If anyone symbolized what it meant to "be scientific" and "do science" during the Victorian period, it was John Frederick William Herschel. Born on March 7, 1792, into astronomical royalty as the only son of William Herschel (1738–1822) and Mary Baldwin-Pitt (1750–1832), John was heir apparent to a scientific legacy that included the construction of immense telescopes and the discovery of hundreds of new celestial objects – including the solar system's first new planet, Uranus. Yet John would prove to have an even longer, broader, and more influential career than his famous father. During a lifetime that spanned Britain's entire age of reform, he achieved fame as a mathematician, chemist, optical experimentalist, astronomer, and public spokesperson of science. From his early contributions importing and developing advanced Continental mathematics into England, to laying the chemical foundations of photographic development with his discovery of hyposulfite "fixer," to extending and establishing the pioneering work of his father and his aunt, Caroline Herschel (1750–1848), in sidereal astronomy, John Herschel came to embody the pursuit of natural knowledge in the nineteenth century.

More than just a keen experimenter and observer, Herschel shaped the practice – and indeed the very meaning of – science during this period. Throughout his life, he was at the center of an immense network of scientific correspondence, stretching throughout Europe and beyond, that linked scientific practitioners in North America and the far-flung reaches of the British Empire. As first foreign secretary of the Astronomical Society of London (soon to become the Royal

Astronomical Society) and as secretary to the Royal Society of London, Herschel worked to reorganize British science and standardize scientific observation and measurement around the world. He attempted to make the scientific establishment less aristocratic (and, by extension, less conservative in mind and spirit) by cofounding new learned societies and allowing himself to stand for election as president of the Royal Society against the king's own brother. From an acute awareness that science needed data, which, in turn, demanded more practitioners, Herschel encouraged voyages of discovery and immense geophysical endeavors such as the "Magnetic Crusade" and collaborated with the British Admiralty to ensure sailors were trained as accurate scientific observers, casting an observational net for science around the globe.[1] This impetus grew from his own early travels throughout England and then Continental Europe, measuring the Alps with thermometers, barometers, and instruments of his own design; collecting chemical and mineralogical specimens from the slopes of Etna and Vesuvius; and standardizing, testing, and comparing equipment in astronomical observatories from Sicily to Göttingen. Herschel's voyages culminated with a four-year scientific pilgrimage to the British colony of Cape Town, in South Africa, where, by completing his astronomical surveys of the southern skies, he became the first astronomer in history to telescopically survey the entire heavens.[2] Yet even at the Cape his time included measuring, collecting, and sketching terrestrial as well as celestial objects.

Herschel's drive to quantify and standardize his observations extended from the position of double stars or the appearance of nebulae at the observational limits of astronomy to careful camera lucida drawings of architectural and geological vistas. He was a draughtsman in the final generation before photography was applied to both terrestrial and astronomical scenes. Indeed, his career spanned this transition, symbolically encapsulated by the photographic exposures he made of the framework of his father's giant forty-foot telescope. Whether it was the ruins of the Roman coliseum or the strata of an Alpine valley, Herschel's measurements were informed by an artistic eye and trained hand.[3] Self-consciously fashioning himself as the all-around man of science, Herschel came to embody the Romantic (Humboldtian) appeal of science, unifying an artistic and literary disposition with obsessive quantification.

Herschel could exert such influence because of the leadership roles he played within the scientific community. Educated privately – interrupted only by a brief stint at Eton at the age of eight – Herschel entered St. John's College, Cambridge, in 1809, where his abilities soon became apparent. In 1813, at the age of twenty-one, he graduated Senior Wrangler, became First Smith's Prizeman, and was elected as a fellow of his college and of the Royal Society. At Cambridge, he met Charles Babbage (1791–1871) and George Peacock (1791–1858), with whom he founded the Analytical Society. Upon graduation, Herschel had planned for a career in law, entering Lincoln's Inn in 1814, but he returned to Cambridge that same year to work as mathematics tutor, which he disliked. At twenty-three, Herschel stood as a candidate for the Chair of Chemistry at Cambridge, though withdrawing from competition before it went to vote. In 1816, he began to study astronomy and decided to leave Cambridge to continue his father's astronomical observations. A few years later, in 1821, he cofounded the Astronomical Society, which in 1831 became the Royal Astronomical Society, of which he was president in 1827, 1839, and 1847.

As a founding member of both the Analytical Society and the Astronomical Society, Herschel was instrumental in advocating new practices and promoting a cautious reforming spirit in science.[4] (Herschel disliked Babbage's more "violent," declinist approach.) The Analytical Society was self-consciously revolutionary in its mathematical pursuits, promoting non-Newtonian, non-geometric algebra at Cambridge, that bastion of Newtonianism, whereas the Astronomical Society was perceived as a direct threat to the hegemony of the venerable Royal Society.[5] In the Astronomical Society, Herschel's leadership created a space not only for applying new mathematical methods to astronomy but for middle-class bankers, stockbrokers, and school masters to engage science outside the orbit of the aristocratic Royal Society. Herschel also played a leadership role within more established scientific circles, though always with a reforming drive away from traditions of privilege and toward meritocracy. This culminated with his bid for Royal Society presidency, a scandal in itself as he took the audacious step of running against the king's brother, the duke of Sussex. His defeat by the Duke did little to decrease his scientific prestige, which was only strengthened by his time away from London at the Cape of Good Hope. In Cape Town, Herschel was instrumental in reforming educational practices in the colony, and upon his return to England he participated in the newly formed British Association for the Advancement of Science. Late in his life, like Newton before him, Herschel served as Master of the Mint, where his reforming tendencies continued as he swept away generations of privileged inefficiencies and worked to rationalize and standardize British currency.[6] Throughout his later career, his influence was exerted through his correspondence, his suggestions to other researchers, and his published writings, including popular textbooks and lectures.

This influence extended far beyond the scientific community through Herschel's publications as what would now be termed a "best-selling" author. His initial mathematical papers had little influence outside the mathematical community; at the same time, his translation (with Babbage and Peacock) of the algebra textbook of the French algebraist Sylvestre Lacroix (1765–1843) provided the impetus for eventual mathematical reform at Cambridge by providing the first accessible resource for teaching and learning Continental analysis. Indeed, in the *Dictionary of National Biography* these works of Babbage, Peacock, and Herschel on "differential notation" and "continental methods of analysis" were said to have been responsible for "the restoration of mathematical science in England."[7] Herschel went on to publish long encyclopedia articles on "Sound," "Heat," "Physical Astronomy," and "Light." These – most especially "Light" – provided technical surveys of each field and showcased his extensive reading and correspondence. "Light" was quickly translated into French and German and considered the most important work of optics after Newton. His 1833 *Treatise on Astronomy* (eventually expanded to the *Outlines of Astronomy*) went on to become one of the most popular astronomy texts of the nineteenth century, going through multiple editions and being translated into many languages, including Chinese. People who learned their astronomy during this period (including other astronomers and popular astronomy writers) learned it from the *Treatise/Outlines*. Though Herschel's father may have brought an awareness of the universe beyond the solar system to other natural philosophers, it was Herschel, through his popular writings, who brought an awareness of this "sidereal revolution" to the general public.

For enduring importance to the history and philosophy of science, however, none of Herschel's works were as influential as his *Preliminary Discourse on the Study of Natural Philosophy*. Written as an introductory volume to a series on science within a larger encyclopedia series and drafted in the months leading up to his bid for presidency of the Royal Society, the *Preliminary Discourse* offered a survey of the state of the physical sciences, an apology and appeal for a life devoted to the pursuit of science, and an outline of the methods of science for a popular audience. The rules for scientific investigation that Herschel laid down in the second portion of this work were the first attempt since Francis Bacon (1561–1626) to systematically show how discoveries were made, and the contents were given immense credence due to their author writing from his own extensive experience. Herschel's *Preliminary Discourse* was the starting point of the more elaborate systems of inductive philosophy of science later developed by William Whewell (1794–1866) and John Stuart Mill (1806–73), though some commentators since have considered the book as an early articulation of the modern hypothetico-deductive method.[8] More recent analyses of the *Preliminary Discourse* have cast it as a "conduct manual," teaching its middle-class readers to better themselves through science, and another example of Herschel's reforming tendency, a bid to move science from the realm of aristocratic privilege by unfolding its methods to all.[9] There is no doubt that the book was a sensation in its day, providing a compelling vision of the scientific life and a model for how science was done and influencing other practitioners such as Michael Faraday (1791–1867) and Charles Darwin (1809–82).[10] Widely read and excerpted, the *Preliminary Discourse* established Herschel as a spokesman for science – with the science advocated being not simply a method of investigating the world but a way of *being* scientific, linked to theistic piety and the epistemic virtues of patience, disinterest, and humility.

After Herschel's return from the Cape of Good Hope in 1838, his scientific output slowed even as his influence grew. He was first knighted for his service to science and ultimately created a baronet (a hereditary order of minor nobility that passed to his eldest son) by Queen Victoria. At the time of his death, Herschel had recently completed a general catalog of all the nebulae discovered by him and his father (which forms the basis of the *New General Catalogue* still used for deep sky objects today) and was at work on a similar catalog of all double stars. At the end of his career, he had returned to his very first astronomical project: making the observations of himself and his father accessible and useful as an empirical baseline for other observers for generations to come. Upon his death on May 11, 1871, Herschel was buried with honors in Westminster Abby, with an inscription that alludes to the esteem in which he was held in both popular and scientific perception:

NATU OPERE FAMA FILIUS UNICUS "COELIS EXPLORATIS" HIC PROPE NEWTONUM REQUIESCIT

GENERATIO ET GENERATIO MIRABILIA DEI NARRABUNT

The only son of the famous "heaven explorer" rests here near Newton;
Generation to generation proclaim the wonders of God

Herschel Scholarship: Challenges and Prospects

The incredible breadth of Herschel's work, the vast amount of existing manuscript resources for exploring it, the length of his life and extent of his influence all have made Herschel a difficult figure to reduce to a single book or study.[11] Though his published books and articles are readily available, the sheer volume of letters written during his lifetime (over 14,000 surviving) and the number of unpublished manuscripts, notebooks, diaries, drawings, and miscellaneous jottings make a complete treatment of his life and career daunting. The most important resource to the Herschel scholar to date remains *A Calendar of the Correspondence of Sir John Herschel* (Cambridge University Press, 1998), edited by Michael J. Crowe, David R. Dyck, and James R. Kevin. This reference provides a chronological listing of almost all of Herschel's surviving correspondence as well as short summaries for each letter, along with a complete bibliography of

Herschel's works as well as works on Herschel up to the time of publication. An online, searchable version of the *Calendar* is hosted by the Adler Planetarium and also available through Epsilon.[12] The majority of these letters are held by the Royal Society, the Royal Astronomical Society, and the Harry Ransom Center at the University of Texas, Austin, though dozens of other repositories hold smaller numbers. Herschel's papers are mainly divided between the Royal Astronomical Society, the Science Museum (where his chemical notebooks have been scanned and can be perused online), the Harry Ransom Center, and the Herschel Family Archive, which is maintained by Herschel's heirs.[13]

This volume draws on these and other resources to bring together for the first time introductory surveys of all major aspects of Herschel's life and work. The chapters in this *Cambridge Companion to John Herschel* offer an unprecedented view of the breadth and depth of Herschel's myriad scientific activities, from his mathematical work to his photography, from his drawing practice to his philosophy. In selecting these chapters, we chose to focus on the central themes of Herschel's scientific practice, output, and legacy, and in such a way that other themes and topics (outlined below) could flow naturally from them. Taken together, they provide the student or general reader with an introduction to the key aspects of Herschel's life and influence. Collectively, the chapters offer a perspective on Herschel that emphasizes his role in standardization and quantification across the natural sciences but complicates this picture significantly. In particular, though Herschel has often been seen as part of a joint mission with Cambridge colleagues like Babbage and Whewell to reform science, such narratives often gloss what was specific to Herschel himself.[14] These chapters draw attention to the ways in which Herschel carved out his place in the wider intellectual landscape of the period and differed from other major figures of the time.

John Herschel is an especially critical figure in the history of nineteenth-century science, and the chapters in this volume should be seen as opening avenues for future research. Herschel, for instance, was *the* leading scientific figure during a time of significant transition as the practice of natural philosophy, with its assumption of a creator and emphasis on personal virtue, became the practice of modern science, dominated by mathematics with a hierarchy based on mathematical and experimental skill rather than social class. His work – as well as his interactions with colleagues like Mary Somerville (1780–1872), David Brewster (1781–1868), Charles Babbage, and William Whewell and their competing visions of science – provides an avenue for understanding this change and how it was perceived, even as he resisted aspects of it such as professionalization and the division of science into separate disciplines.[15] Likewise, Herschel provides an important case in the interaction between science and faith during a period of intense and evolving piety in England.[16] Through his marriage to Margaret Brodie Stewart (1810–84), with whom he had twelve children, Herschel became linked to an evangelical network centered on Marylebone Church, London, that included such important figures in the evangelical movement as Charles Wesley Jr. (1757–1834). How did Herschel's faith influence his role in the transformation from natural philosophy to modern science?

As another example, Herschel's correspondence networks show the interactions during this period between the government and science practitioners, as Herschel worked to balance keeping science independent of government while simultaneously linked with it through networks of support and influence. Herschel is central to understanding the role of the popularization of science during this period as well, especially in exploring how writing for different audiences influenced the expression of science.[17] His output provides a nearly complete cross-section of the type of science writings produced during this period: from very technical mathematical papers and encyclopedia articles to books written for the widest audiences. Herschel's relationship with his public and his various publishers and editors are a fruitful means for understanding the history of writing and reading publics as well as the history of the book in Victorian England. In addition, Herschel's time as secretary of the Royal Society and his leadership roles in the Astronomical Society meant he was instrumental in establishing the practices of modern scientific communication, including peer review of scientific papers and, in the case of the Astronomical Society, establishing the tone and procedures for what would appear in its journal.[18] Further work remains to be done examining Herschel's role in the relationship between science and government, scientific popularization during this period, and the formalization of scientific communication.

There is also much to be done on Herschel and aspects of imperial science, on the role of colonization during his time in South Africa, its influence on his work, and astronomy as a colonizing science – to say nothing of his magnetic and tidal surveys and his promotion of science through the reach of the British navy.[19] His work with Margaret on the flora of South Africa also remains largely unexamined, as well as the role he played in promoting and supporting women in science, such as his aunt but also the science writer Mary Somerville and the early photographic pioneer Julia Margaret Cameron (1815–79). Finally, there is the role Herschel played in defining the identity of science – the traits and values of the scientific practitioner, the perception of what the scientist is and does, and *where* science is conducted. (Herschel, like Darwin, is an example of science in the domestic setting.)[20] Herschel was central in establishing scientific identity during the period in which the term "scientist" was only beginning to replace "natural philosopher."[21] It is our hope that the chapters in this volume lay the groundwork for future research on these topics.

There is certainly danger in romanticizing the heroic person of science – and Herschel lived a life that provided a template to do just that. He could be viewed as the "last polymath," with one foot in the traditional world of natural philosophy and the other in the new world of modern science that he helped shape and define, even as he also gained recognition for his literary and artistic endeavors, including translations of the *Iliad* and German Romantic poets. Yet if Herschel scholars seem prone to such romanticization, part of the reason is surely spending so much time in the letters and papers of someone who comes across the centuries as genuinely sympathetic, humble, quick to give credit to others, doting on his wife and many children, and more interested in the quiet pursuits of science and family at home than prestige among the scientific circles of London. We hope the chapters in this volume open aspects of his work to others and provide entrances into the various avenues of study alluded to above – while also communicating something of who Herschel was as a person.

Notes

1 John Cawood, "The Magnetic Crusade: Science and Politics in Early Victorian Britain," *Isis*, 70.4 (1979): 493–518; John Herschel, ed., *A Manual of Scientific Inquiry: Prepared for the Use of Her Majesty's Navy and Adapted for Travelers in General* (London: John Murray, 1849).

2 See Chapters 3 and 4 of this volume.

3 See Chapter 6 of this volume as well as Larry Schaaf, *Tracings of Light: Sir John Herschel & the Camera Lucida* (San Francisco: The Friends of Photography, 1989); and Omar W. Nasim, *Observing by Hand: Sketching the Nebulae in the Nineteenth Century* (Chicago: University of Chicago Press, 2013).

4 On political style and scientific reform in the early Victorian era, see, for instance, Joe Bord, *Science and Whig Manners: Science and Political Style in Britain, 1790–1850* (New York: Palgrave Macmillan, 2009).

5 See Chapter 2 of this volume.

6 See Chapter 9 of this volume.

7 Agnes Mary Clerke, "Herschel, John Frederick William," in *Dictionary of National Biography* (London: Smith, Elder & Co., 1885–1900), vol. 11, 263.

8 See Chapter 5 of this volume.

9 James A. Secord, "The Conduct of Everyday Life: John Herschel's *Preliminary Discourse on the Study of Natural Philosophy*," in *Visions of Science: Books and Readers at the Dawn of the Victorian Age* (Chicago: University of Chicago Press, 2014), 80–106. For the latter view, see Stephen Case, *Creatures of Reason: John Herschel and the Invention of Science* (Pittsburgh: University of Pittsburgh Press, 2024).

10 See Chapter 11 of this volume.

11 The only biographies of Herschel remain Günther Buttmann, *The Shadow of the Telescope: A Biography of John Herschel* (Guildford: Lutterworth Press, 1974); and the privately published Eileen Shoreland, *Sir John F. W. Herschel: The Forgotten Philosopher* (The Herschel Family Archive, 2016). Case's forthcoming *Creatures of Reason* will provide a detailed biographical study of Herschel's life from his time at Cambridge to the publication of the *Preliminary Discourse*.

12 These letters are searchable online at http://historydb.adlerplanetarium.org/herschel/ and https://epsilon.ac.uk. Another important resource that deserves mention here is Brian Warner, ed., *John Herschel 1792–1992: Bicentennial Symposium* (Cape Town: Royal Society of South Africa, 1994) – published in the Society's *Transactions*.

13 Scans of Herschel's experimental notebooks held by the Science Museum can be accessed at https://collection.sciencemuseumgroup.org.uk/documents/aa110066295/collection-of-john-herschels-notebooks.

14 See, for instance, Laura J. Snyder, *The Philosophical Breakfast Club: Four Remarkable Friends who Transformed Science and Changed the World* (New York: Broadway, 2011).

15 Snyder, writing in the tradition of Walter (later Susan Faye) Cannon, has emphasized the commonalities between Herschel and men like Babbage and Whewell. See Snyder, *The Philosophical Breakfast Club*. For a contrasting perspective see, for instance, William J. Ashworth, *The Trinity Circle: Anxiety, Intelligence, and Knowledge Production in Nineteenth-Century England* (Pittsburgh: University of Pittsburgh Press, 2021); and Chapter 10 of this volume. For essays on aspects of the theme of professionalization see, for instance, Bernard Lightman, ed., *Victorian Science in Context* (Chicago; University of Chicago Press, 1997); and Bernard Lightman and Bennett Zon, eds., *Victorian Culture and the Origin of Disciplines* (New York; Routledge, 2020).

16 See, for instance, Richard Bellon, *A Sincere and Teachable Heart: Self-Denying Virtue in British Intellectual Life, 1736–1859* (Leiden: Brill, 2015); and Jonathan R. Topham, *Reading the Book of Nature: How Eight Best Sellers Reconnected Christianity and Science on the Eve of the Victorian Age* (Chicago: University of Chicago Press, 2022).

17 See, for instance, Geoffrey Cantor and Sally Shuttleworth, eds., *Science Serialized: Representation of the Sciences in Nineteenth-Century Periodicals* (Cambridge, MA: The MIT Press, 2004); and Bernard Lightman, *Victorian Popularizers of Science: Designing Nature for New Audiences* (Chicago: University of Chicago Press, 2007).

18 See Alex Csiszar, *The Scientific Journal: Authorship and the Politics of Knowledge in the Nineteenth Century* (Chicago: University of Chicago Press, 2018); and, especially, Alex Csiszar, "Proceedings and the Public: How a Commercial Genre Transformed Science," in Gowan Dawson, Bernard Lightman, Sally Shuttleworth, and Jonathan R. Topham, eds., *Science Periodicals in Nineteenth-Century Britain: Constructing Scientific Communities* (Chicago: University of Chicago Press, 2020), 103–34.

19 Elizabeth Green Musselman, "Swords into Ploughshares: John Herschel's Progressive View of Astronomy and Imperial Governance," *The British Journal for the History of Science*, 31.4 (1998): 419–35.

20 See, for instance, David N. Livingstone and Charles W. J. Withers, eds., *Geographies of Nineteenth-Century Science* (Chicago: University of Chicago Press, 2011); Donald L. Opitz, Steffan Bergwik, and Brigitte Van Tiggelen, eds., *Domesticity in the Making of Modern Science* (New York: Palgrave Macmillan, 2016).

21 The place to begin with this remains Sydney Ross, "*Scientist*: The Story of a Word," *Annals of Science* 18.2 (1962): 65–85.

1
———————

John Herschel
A Biographical Sketch

John Herschel grew up in a world of intense scientific activity. From an early age, he was surrounded by examples of science in action. His education was supplemented at every stage to ensure exposure to and training in the highest, most up-to-date mathematics. At the same time, John also experienced an indirectly political upbringing, one that showed him not only a changing world but also the power of education to challenge and change tradition. The politics he witnessed growing up illustrated dramatic transformations taking place in England and abroad. Family friends were caught up in the aftermath of the French Revolution. In England, family holidays and changing consumer tastes evidenced the effects of England's industrial revolution. In adulthood, these two strands – the scientific and the political – influenced his outlook and choices. He chose friends at university, such as Charles Babbage (1791–1871) and George Peacock (1791–1858), who wanted to change the world through mathematics. He attempted to earn a living from natural philosophy rather than train for the church or fall back on family wealth after graduation as his father wished. He chose to marry an evangelical free thinker, keen to use knowledge and religion to improve the lives of others. He chose to never completely pursue science in isolation but instead advise governments both in South Africa and in England on using science to improve education and society. He chose to train his children to be both scientifically curious and socially minded. These choices shaped Herschel's life.

John's father and aunt, William and Caroline, are well known in the history of science. William discovered the planet Uranus and pioneered the study and surveying of deep-sky objects, including nebulae, star clusters, and double stars. Caroline is known today for her comet discoveries and her work assisting William. Both discovered early on that making one's name in eighteenth-century astronomy was easier when the science was accompanied by a certain style of self-presentation. John's story can be understood in similar terms. His campaigning work was as important as his scientific work in making him known as the archetypical man of science of his day. Whether this was his involvement in political projects aiming to raise the profile of a continental style of mathematics (see Chapter 2) or his promotion of national philosophy as central to mainstream society, these efforts worked with his intellectual achievements. By the 1830s, John Herschel had come to exemplify the British man of science. Artist and critic John Ruskin (1819–1900), for example, could speak of having 'met the leading scientific men of the day, from [John] Herschel downward'.[1] Historian Susan Cannon described the definition of a nineteenth-century man of science as being 'as much like Herschel as possible'.[2] Had John followed his father's wishes – studying hard and joining his father at the telescope straight from university without ever speaking out on political matters or trying to make an independent career for himself in science – he would not have attained his position of scientific influence. In the biographical sketch that follows, I follow John Herschel's path as he created for himself an intricate, complicated, and nuanced life in science.

Childhood

Parents almost always want more for their children than they had themselves, and William Herschel was in this sense a typical parent. Through a careful selection of schools, tutors, and university, William created the perfect scientific education for his son. On the one hand, John was trained to excel in complex forms of mathematics and fit seamlessly into the highest echelons of British society. On the other, he was brought up to see society as both changing and as something possible to change.

John Herschel grew up in Observatory House in Slough, a town near London and just a few miles from the king's residence in Windsor. This was the house his father and aunt had settled in after William had become the king's astronomer, and its grounds held the various large telescopes used for their work in astronomy. Much of what we know about John's early childhood comes from reminiscences of his aunt Caroline.[3] From these, we learn that John copied in play the actions of the practitioners of science by whom he grew up surrounded. When very small, he would visit Caroline as she worked on editing the astronomical catalogue of John Flamsteed (1646–1719) for William's surveys. 'When the Nurse brought him to me', Caroline recounted, 'he came with Aunty shew me the Wail'.[4] Around the age of two, John would follow the workmen building, maintaining, and operating William's telescopes, copying what they did. Caroline described, for example, an incident when 'my attention was drawn to his hamering [*sic*]', and she had to call over a carpenter to fix the damage young John had wrought. When he started school and returned to stay with Caroline during the holidays, he would again play at what he had seen at Observatory House, this time mimicking the work in the chemistry buildings where the telescope mirrors were made. 'Many a half, or whole holy day', she recalled,

> he was allowed to spend with me and dedicated to making experiments in chemistry where generally all boxes, tops of tea canisters, pepper boxes, tea cups &c served for necessary vessels and the sand tub furnished the matter to be analized. I only had to take care to exclude water which would have produced havoc on my carpet.[5]

Children play at being what they see around them, and John saw astronomers, instrument makers, and chemists. Whereas his father had grown up with a musician father and had begun to learn the violin as soon as he was big enough to do so, John grew up surrounded by experimental and observational science before he had even started school. Though this did not make his future life in science inevitable, it did make it easier to imagine.

In the second half of the eighteenth century, a proliferation of private schools and academies were set up across Britain.[6] Little documentation remains for the majority of these schools or their teachers save for advertisements in local newspapers and references in domestic correspondence. They tended to be small, catering to only about ten children at a time, and taught a range of subjects that for boys would generally include writing and arithmetic. John's first school was in Slough and run by a Mr Atkins.[7] He started school at the age of five; yet despite his youth, his father employed an additional tutor for him in writing, arithmetic, and geography so John could keep studying on his days off. Next, John attended a Mr Bull's school in Newbury (around thirty miles from Slough) and, by August 1799 (at the age of seven), was able to write letters home to his mother characterised by good spelling, neat penmanship, and careful grammar.[8]

A year later, John was sent to Eton, close to Slough. This was a prestigious step in his education, and his parents commissioned a portrait of young John in front of the school gates to honour his enrolment.[9] John's mother, Mary Pitt Herschel (1750–1832), was a local woman whose first husband and son had both gone to Eton.[10] Other members of her family may have been educated there too, and many of them went on to attend Oxford and Cambridge to become doctors, lawyers, and clergymen.[11] John's early education was thus shaped by this family tradition on his mother's side. It was an upper-middle-class education designed to prepare boys to run family estates or enter one of the professions.[12] In most families this would have been enough – but not for William.

Three weeks into John's time at Eton, his parents removed him from the school. Caroline tells a story of Mary seeing (and being horrified by) John fighting. Later, she also mentioned William complaining that at Eton, boys 'were kept with learning nothing but Greac [sic] and Latin till the age of 18 which he called a waste of many years of the life of a young man'.[13] Whatever the catalysing incident, these shortcomings likely contributed to John's quick removal and the beginning of a new, carefully designed education. After only three weeks at Eton, John moved to a school run by George Holdsworth Lowther Gretton (1780–1833).[14] The school was in Hitcham, Buckinghamshire, about five miles from Slough, and, according to Caroline, Dr Gretton did nothing without William's 'approbation and advice'.[15] Though this level of control and involvement was enough for the first few years at Dr Gretton's, by 1804 (when John was twelve) William decided to supplement his mathematics instruction with a home tutor, a recent Cambridge graduate named Matthew Henry Thornhill Luscombe

(1776–1846). Two years later, when John was fourteen, William replaced this tutor with Alexander Rogers (fl. 1807–33). Rogers was Scottish and, like many educated in Scotland rather than England, had some understanding of both Continental calculus and its English equivalent, Newtonian fluxions. This was particularly significant for William, because it was this area of mathematics that his French colleagues had used to calculate the orbit of Uranus but that he himself had never confidently understood. Two years after this, a Dr Hausman was also appointed to help John with his German.[16]

Mary, meanwhile, had her own ambitions for her son. In 1792, her son from her first marriage, Paul Adee, died shortly after John was born. Paul was nineteen years old and only recently out of Eton. This may help explain her concerns regarding Eton and her later choice to stay near John when he first moved to Cambridge. As John got older, Mary's experience and opinions began to play a greater role in shaping his education. It was from Mary's knowledge and local contacts that the various local schools John attended were chosen. It was Mary, rather than William, who had prior knowledge and connections with Eton. Though the curriculum and additional tutors were William's domain, Mary had local knowledge and family tradition to shape John's path.

One major change Mary introduced to the Herschel household was to insist on regular travel when she married William. Each year, William, Mary, and John – often accompanied by a cousin – would travel to visit friends and places of interest. Through these travels, John glimpsed the changing world around him. When John was ten, for example, the Herschels went on a sightseeing tour of post-revolutionary Paris. Later, when he was in his late teens, they took a series of tours of the changing landscape of industrialising Britain. John and his mother's accounts of these show their shared love of romantic poetry and of the natural beauty of the Lake District.[17] They also show their conflicted reaction to industrialisation. The silk and cotton mills in Derby, for example, were fascinating due to 'the elegant machinery employed'.[18] In Grantham, a growing town in Lincolnshire, Mary was impressed with the efficiency of the steam engines and threshing machine.[19] However, they were concerned by the high density of manufactories in Manchester and the hording of wealth by the factory owners. Of Richard Arkwright Junior (1755–1843), referred to in Mary's travel

journal as owning 'several of the cotton manufactorys in this neigh-
bourhood', Mary wrote that his income of £100,000 per year was 'too
much for any one man'.[20] By 1799, thanks to his mills and investments,
Arkwright was one of Britain's ten millionaires. John would soon feel
that the nation as a whole was 'choked with its own population –
overloaded with its useless manufactures & decaying commerce'.[21]

From 1810 onwards, the Herschels concluded their holidays in
Scotland, visiting Mr Rogers and some of William's friends. It was here
that John met his lifelong friend James Grahame (1790–1842), a law
student at Glasgow University. Grahame remained one of John's
closest, longest-standing friends and perhaps his only close friend not
involved in mathematics or science. While visiting, John would accompany Grahame to balls and parties, and Grahame would try to get him
to read the novels of Jane Austen. For decades, their shared correspondence was filled with stories about their lives, loves, aspirations, and
deepest desires. As John told his future mother-in-law when introducing Grahame as a character witness, 'I have no secrets, or intentional
concealments of any kind from my noble-minded friend – that which
I could not communicate to him, I would not harbour in my own
bosom'.[22]

Cambridge

In 1810, John began his studies at St John's College, Cambridge, where
he was drawn to a group of high-achieving middle-class students set on
revolutionizing British mathematics. How the Herschels came to
choose St John's is unclear, though it is possible it was not natural
philosophy that determined the choice but rather politics and religion.
At the time, St John's was known to be an anti-slavery, abolitionist
college, a position that would have been applauded by John's mother
who lived in rooms down the road during John's first term at
Cambridge.[23] During the holidays, his Cambridge tutor was invited to
Slough to be interrogated by his parents.[24] John, meanwhile, wrote
frustrated letters to his old tutor, Mr Rogers, complaining that he was
bored and none of his lectures were teaching him anything new.[25]
To keep busy, John began writing. Just as his father had made his
writing debut in the popular magazine *The Ladies' Diary*, John started

his writing journey with an anonymous letter to the popular *Nicholson's Journal* (more properly known as *A Journal of Natural Philosophy, Chemistry and the Arts*) about a new method for deriving various formulae for sine and cosine.[26]

Studying at Cambridge in the early 1800s was not entirely the experience the Herschels had imagined. Their knowledge of Cambridge came mainly from their high-ranking scientific peers – friends like William Watson (1744–1824) and Nevil Maskelyne (1732–1811) – who had gone to Cambridge and who were now well respected for their expertise in mathematics and science. Watson, like John, was the son of a successful man of science and was at Cambridge in the 1760s. Maskelyne, the Astronomer Royal, had been at Cambridge even earlier, in the 1750s, so their experience as students was fairly out of date. Both Watson and Maskelyne, however, inhabited a world where studying at Cambridge was a common route to fitting in within the Royal Society and English natural philosopher life. The Herschels were part of that same world by the time John was born, so for William at least, his knowledge of Cambridge graduates was people like Watson and Maskelyne rather than the sons of country squires that made up a greater proportion of Cambridge undergraduates at the time. The Herschels were unaware that academic excellence was the exception rather than the rule for the majority of Cambridge students.

Cambridge in the early 1800s was one of a very small number of universities in the United Kingdom and existed primarily to train the eldest sons of aristocratic families to run their family estates, and younger sons to enter the professions, namely medicine, law, or the church. Most students were aristocrats, but there were also a few from the new middle classes, including those, like the Herschels, who had used education to raise their social status. There was also at this time a proliferation of student societies.[27] Students could join anything from the Bible Reading Society to a variety of drinking clubs depending on their interests. Two students, Edward Bromhead (1789–1855) and Charles Babbage (1791–1871), found the efflorescence of societies ridiculous. Looking at the aim of the British and Foreign Bible Society to distribute Bibles, they decided to call a meeting of their own society, summoning 'all those of his acquaintance who were most attached to mathematical subjects' to a meeting in May 1812 (towards the end of

John's second year at Cambridge) to discuss the distribution of the calculus textbook of Sylvestre François Lacroix (1765–1843).[28] The attendees of this meeting formed the Analytical Society and became John's friends and collaborators for many years. They included Babbage, George Peacock (1791–1858), John William Whittaker (1790–1854), and Richard Gwatkin (1791–1879). All were atypical for Cambridge undergraduates, interested in studying not just English fluxions (a standard part of the Cambridge syllabus) but French calculus as well (which was most definitely not) (see Chapter 2).

Though mathematics was their primary interest, John, Babbage, and their friend William Whewell (1794–1866) were also interested in philosophy more broadly and with a political as well as an intellectual agenda. To discuss this interest, they would meet with Richard Jones (1791–1855) every Sunday for breakfast in what historian Laura Snyder has termed the Philosophical Breakfast Club.[29] In these meetings, they would eat a huge breakfast and then sit around the fire discussing a chosen topic and generally putting the world to rights.

Groups like the Breakfast Club and the Analytical Society helped John forge his identity as a man of science, giving him a circle of like-minded friends and a sense of what he could and should accomplish with his education. Upon graduation, his father assumed John would train for the church, take a parish, and become a leisured gentleman of science, working with William on their joint astronomical enterprise. John, however, saw a different path for himself. Many of his Cambridge friends were training for professions, and John decided he would do the same. He first attempted to become a lawyer, like his friend James Grahame. When he found work in a law office incredibly boring, he tried to earn a living as a chemist or mineralogist. Only when he discovered these jobs were rare and hard to secure did he begin to come round to his father's initial proposal. After leaving Cambridge, John dabbled in chemistry, mineralogy, and physics, experimenting and seeking out mentors and teachers throughout Europe. Gradually, he began to form an increasingly coherent view of science. He learned as well through these new relationships and experiences that the intellectual pursuit of mathematics and natural philosophy could be about more than intellectual curiosity, personal advancement, and building a reputation. It could be about making the world a better place.

Beyond Cambridge

Between graduating in 1813 and returning to Slough in 1816, John tried and failed to find work he enjoyed. He moved to London to study law, but instead attended meetings at the Royal Society (a possibility thanks to his father nominating him as a fellow) and, from what he learned there, began to teach himself experimental chemistry. In September 1814, he returned to Cambridge (he had been a senior wrangler and was awarded the Smith's Prize and a fellowship for five years), where he began attending lectures in mineralogy and found a mentor in Professor Edward Daniel Clarke (1769–1822). At the same time, he and Babbage exchanged letters sharing both their progress in these new experimental sciences and their failures in the job market. Babbage suggested John apply for the post of professor of chemistry at Cambridge, though John's experience as an assistant tutor convinced him that any post involving any teaching was not for him.

One area of work did seem to interest John, though – writing. Babbage was not so keen on the idea, as he had a family by this point and wanted a job with a regular salary. John had more flexibility. As he told his friend:

> I mean to publish ... I hear you have been writing papers for this new journal of the Royal Institution. Who are the editors of it? Will they pay one for writing? Or do you know anybody who will? I should like to write articles for some of these encyclopaedias if I could . . .. In a word I want to get money and reputation at the same time, and to give up cramming pupils, which is a bore & does no one credit but very much the contrary.[30]

In 1816, William Herschel became ill and asked his son to return home. John obliged and began the long process of learning to become an instrument maker and observer. Like Caroline, John was given a journal in which to record his 'sweeps'. His father also took him through the process of rebuilding the twenty-foot reflector, the primary telescope William used for his nebulae surveys. Gradually John became interested in and knowledgeable about astronomy. At the same time, he continued his research in other areas. He corresponded with Clarke, his mineralogy mentor in Cambridge, learning different ways of analysing minerals and sometimes joining Clarke at scientific gatherings, where he no doubt enjoyed being introduced as Clarke's mentee rather than William Herschel's son. As his expertise grew, John began to explore connections between disciplines. He used his chemical knowledge in analyses of mineralogical specimens. He combined this with optics, looking at polarisation and other optical properties of crystals. By 1820, John had started publishing his findings (until then all his publications had been in mathematics), building his reputation as a man of science and not just a mathematician and astronomer.

For the next few years, John supplemented his training and research at home with travels and collaborations abroad. His first trip was to Paris with Babbage in 1819. There, he introduced himself to Jean-Baptiste Biot (1774–1862), an expert in polarisation. When he returned home, John wrote to Biot requesting information for use in his article on polarisation.[31] In August 1821, John and Babbage ventured into the Alps, travelling through France, Italy, Germany, and Switzerland. On this trip, the two young natural philosophers took apparatus with them and made barometric readings at key geological sites (see Chapter 8). These collaborations allowed John to train himself in fields other than mathematics and astronomy and so become distinct from his family and their specialisms.

In the field of astronomy, too, he carved out a distinct niche, separate from that of his famous relatives. In 1820, he and several predominantly Cambridge men formed the Astronomical (later Royal Astronomical) Society; and in 1827, John was elected their president. A year later, he began a collaboration with James South (1785–1867) cataloguing double stars, a project that complemented but was separate from the work he was doing with his father (see Chapter 3). The aims of the Astronomical Society fit with John's earlier concerns about finding intellectually fulfilling work. As William Ashworth has noted, the society was set up to allow '"intellectually" competent men of science'[32] to reduce, classify, and interpret astronomical data sent to them by observers around the world.

In 1822, John travelled again, this time with James Grahame to the Netherlands. The trip meant he was away from home when his father died. Two years later, he travelled once more, accompanied by his servant James Child. The two travelled through France, Italy, and Germany, visiting astronomers and observatories. Following his father's death and the foundation of the Astronomical Society, John was now in a different position when introducing himself to astronomers abroad. Previously, Herschel had worked to separate himself from his father's work, emphasising his interest in other fields. Now he was more comfortable allowing astronomy to play a central role, aware that he would no longer be seen as just his father's son.

By the mid-1820s, John had published in a range of scientific disciplines and built contacts through his experimental work in Europe and his involvement with the Astronomical Society. He had also started to write for popular audiences, including the encyclopaedias he had mentioned to Babbage years before. His work building and cultivating a unique, scientifically coherent identity culminated at the start of the 1830s with his publication of *A Preliminary Discourse on the Study of Natural Philosophy*. This scientific bestseller helped shape how future generations would see science and perhaps more importantly, how they would see the utility of science to society (see Chapters 5 and 11).[33] The book began as an encyclopaedia volume commissioned by Dionysius Lardner (1793–1859) for the *Cabinet Cyclopaedia* in 1828. In his request, Lardner explained the difficulty of finding writers who could write for a popular audience and had sufficient subject knowledge – but in John he had found both.

The *Preliminary Discourse* helped define science as a discipline. In it, John drew on his knowledge of chemistry, mineralogy, optics, and astronomy, all areas he had spent the past seventeen years studying and developing. Throughout, he argued for the utility of science to society, science's role as an expression of human curiosity, and scientific investigation leading to tangible utilitarian benefits that enriched lives, brought out the best in humanity, and made society more civilised. He also drew on his experiences with the Astronomical Society and that Society's division between competent men of science and amateur fact-gathers, arguing that some science was 'highly desirable in general education, if not indispensably necessary', whereas the creation of new knowledge

> requires in many cases a degree of knowledge of mathematics and geometry altogether unattainable by the generality of mankind, who have not the leisure, even if they all had the capacity, to enter into such

enquiries, some of which are indeed of that degree of difficulty that they can be only successfully prosecuted by persons who devote to them their whole attention, and make them the serious business of their lives.[34]

When John had first entered Cambridge, he had arrived with mathematical knowledge but no real idea of how to apply it. His father had plans, but through the friends he made in those early Cambridge years John developed a different plan for himself. He found a use for knowledge: not just to improve himself or the total sum of human knowledge but to transform society for the better. These ideas stayed with him and evolved as he and Babbage searched for ways to make a career in science. Through the Astronomical Society, he began to expand this vision to consider how making scientific work and practitioners more central to society might benefit all. This thinking culminated in his *Preliminary Discourse*, where he set out his case and in the process his reputation.

Education and the Cape

John Herschel and Margaret Brodie Stewart (1810–84) met in 1828 when Margaret was in her teens and John was in his thirties. The historian Janet Browne wrote of Charles Darwin's marriage that clever women made him uneasy and so he chose his cousin Emma Wedgwood. John in contrast appears to have been much more at ease with clever women and had good relationships with intellectually accomplished peers such as Mary Somerville (1780–1872) and Maria Edgeworth (1768–1849).[35] His decision to marry an eighteen-year-old as he approached forty, however, suggests he was not looking for 'a companionate wife' to share his intellectual labour or form a 'scientific partnership', so much as he wanted to start a family.[36]

In early December 1828, John proposed, and they married on 3 March 1829. Marriage had a huge impact on John's life and work. The couple had twelve children in all – the first born in 1830, the last in 1855 – which meant a busy family home almost from the very beginning of their marriage. Margaret also introduced John to a new way of thinking about education and public duty. In addition, marrying into Margaret's family gave John access to a world inhabited by the administrators of the British Empire, and so began a new chapter in his scientific career.

Margaret was from a family heavily involved in the evangelical wing of the Church of Scotland. Her father, grandfathers, older brother, and older half-brother were all ministers within the Church and at least by later generations were radical reformers in it as well, keen to focus on belief as a personal journey of redemption.[37] Discussing belief, for example, Margaret's brother told her humans must approach religion as 'rational & immortal beings' and that 'unbelief is an unpardonable sin, other sins maybe forgiven but this never will'.[38] Margaret's family was also well connected to the East India Company. After her father died, Eneas Mackintosh, a London-based merchant with connections to the East India Company, became her guardian. Though two of her brothers were ministers in the Church of Scotland, two more went to India: Duncan Stewart (1804–75) worked for the East India Company as a surgeon, while James Calder Stewart (1806–69) worked for his uncle, James Calder, in his business in Calcutta for a while before moving to Canton.[39]

This background of evangelical Christianity and duty to empire fed into Margaret's family life and childhood education. The education provided by her family (her father, until he died in 1821, and then her mother and older brothers) ended when she married John at the age of eighteen, but up to that point it had included languages (modern and Latin), religion, and some mathematics and astronomy. Margaret was taught astronomy through a religious lens as a means of better understanding the universe created by her God. One of the books her brother set her lessons from, for example, was John Bird's *The Juvenile Observatory*, in which the power and wisdom of God in the works of creation are displayed through a series of astronomical lectures.[40] These lessons made hers an unusually scientific education for an upper middle-class girl, though less surprising as the daughter of a clergyman.[41] Margaret was, according to those who met her,

> a delightful person, with so much simplicity, and so much sense, so fit to sympathise with him [John] in all things intellectual and moral, and making all her guests comfortable and happy without apparent effort.[42]

John meanwhile was as taken with her as he was the warm, nurturing, and affectionate family she came from. 'Were I to choose a family', he wrote to Margaret's mother when asking to marry her daughter, 'I would look to your fireside. Where peace and purity, and innocence seem to have fixed their home'.[43]

John and Margaret's lives changed dramatically within that first year of marriage. Margaret was thrown into a world of highly educated (if kind and welcoming) men and women of science, all much older than her, whom she set about getting to know and learning from. John meanwhile started to read up on parenting to prepare for the arrival of their first child. He ordered Michael Underwood's childcare manual, *Treatise on the Diseases of Children*, and when Margaret had to leave their five-month-old daughter to look after her dying sister, he got up in the night, bottle fed their baby, and kept Margaret informed and reassured of her progress.[44] From the very beginning, marriage and children had a significant impact on John's life and work. He took an active part in raising his children and teaching lessons within their home as they got older. The family had servants as well – a nurse to help care for the babies and a series of governesses and tutors as they got older – but parenting was still a significant change to the life he led before marriage and an impact on the time he had to spare for science.

In 1831, Margaret's brother James wrote to suggest they come and visit him in India. This was the spark that led the family to South Africa. As John explained in his reply to James, his invitation rekindled an earlier desire he had to study the southern skies, and he now was thinking of Cape Town as a potential destination.[45] John had recently failed to be elected as president of the Royal Society, which although not necessarily a personal blow was still a step away from the direction he had hoped British science might go, making time away seem all the more appealing. A year later, in 1832, his mother died, leaving him few remaining familial ties in England. These combined factors led John and Margaret to decide the time was right to travel to South Africa and so complete William's sky survey of nebulae, star clusters, and double stars by mapping the heavens of the southern hemisphere. Cape Town in South Africa seemed a good choice, since it was already home to the Royal Observatory's sister observatory (see Chapter 4). The family – John and Margaret and their children Caroline, Isabella, and William, then aged three, two, and below one respectively – set sail for South Africa in 1833, arriving in early 1834. They stayed for four years, during which time they had several more children (Louisa, Alexander, and John), learned something of colonial life, and played a part in shaping the educational system at the Cape.

While at the Cape, John began to put into practice some of his earlier theories about the role of science in society and education. Soon after their arrival, Rev Dr James Adamson (1797–1875), a Presbyterian minister, financial supporter of missionary work, and key player in the foundation of the South African College, asked if John would suggest a design for a national system of education at the Cape. In response, John first offered a curriculum that included modern and ancient European languages and European history, natural history, the physical sciences, and the useful arts.[46] There was, however, no mention of religion, literature, art, or music. As John learned more about education through both the work of the missionaries who often visited the Herschels at their home of Feldhausen and the education of his own children, he began to revise his views. In his initial response to Adamson, John had said that he felt the purpose of education was to be able to hold government to account and guard against 'obvious & popular fallacies'. At this time, any free man, black or white, who owned property above a certain value could vote.[47] However, the colony still had enslaved people, and they as well as women and free men who did not own property were unable to vote and therefore could not hold their government to account. As John got to know members of these unrepresented groups, his views on education began to change.

Determining the Herschels' attitude to the system of slavery they encountered in South Africa is complex. Margaret described the three women looking after their small children as 'white, free half-coloured, & a coloured slave'.[48] She mentioned it to make a point about her experience contradicting what she had been told, namely that 'white & black – slave & free would not mess together'. How the Herschels came to have an enslaved woman looking after their children is not clear. John's mother had been a strong supporter of abolitionism. Among the few books she owned was one of sermons by the Anglican reformer and opponent of slavery, Beilby Porteus (1731–1809). John's friend Grahame was similarly opposed to slavery, refusing an income from a property in the Caribbean he had inherited through his wife's family since it was farmed by enslaved people. Margaret had been brought up in a strongly evangelical household in which opposition to slavery was an essential part.

Examining how Margaret wrote about her experiences to her family, it seems that where she encountered slavery she viewed it with an intellectual eye, observing and reporting how her observations disproved common misconceptions about enslaved people. At the time, her attitudes seemed liberal; today they come across more condescending. Though she reported proudly that she saw enslaved people as people, she also saw herself and people like her – such as the European missionaries the Herschels came to know – as superior and able to improve the characters of black South Africans. As she told her mother, 'it is impossible not to feel interested in the affairs of the native Hottentots & the slaves'. Later, she wrote to tell her mother that one of her servants ('the one who dresses me') enjoyed Bible stories and 'expresses a great wish to be baptized – for many here live & die without any religious instruction'. Not long after, she described a group of missionary converts as having been 'brought to a very advanced state of civilisation by the Missionaries, for they clothe in the English fashion & have built many brick Houses for themselves'.[49] Margaret was fascinated by the work of the missionaries and brought John to see their work and invited missionaries and their converts to the house. Through them, John began to revise some of his views on education.

Alongside this education in missionaries and their values, John was also learning through the practical experience of working with Margaret to develop an educational curriculum for their own children. To begin with, they employed nurses to help look after their infants. By 1836 (when their eldest was five), they employed one of their nurse's sixteen-year-old daughter as a governess to introduce sewing and multiplication tables. Soon after, they discussed introducing French, music, drawing, and educational walks. They read books on the latest theories on raising children, including Coomb's (or possibly Smiles') *Physical Education* and Maria Edgeworth's *Practical Education*. Both promoted the use of the physical world – walks, gardening, and playing outside – as an integral part of a child's education. Alongside these, Margaret also insisted her children learn their Bible.[50]

As the Herschels prepared to leave South Africa, John once again returned to the question of education. In a letter to Charles Davidson

Bell (1813–82), Surveyor-General in the Cape Colony, he spelled out his revised views on the purpose of education. Education was no longer about teaching European history and languages and sciences with a view to learning critical thinking in order to hold governments to account. Now, he wrote, he felt,

> reading, writing & arithmetic – the perusal of the Scriptures as the foundation of moral instruction and the formation of orderly & moral habits by conformity to a well-regulated distribution of time, occupation & amusement – must form essential features in every plan of education.[51]

In 1828, when he had begun to write his *Preliminary Discourse* and met Margaret, John saw education as preparation for life for a man of property. Within that framework, he saw science education as a means of preparing some men to make science their livelihood and thus make science as important and central to society as law or medicine. He saw a moral element to this, presenting science as a virtuous pursuit as well as one that could be applied to improving people's lives and running society. His view on the transforming power of education continued to expand through his experiences at the Cape. Seeing how missionaries used education and working with his wife to educate his own children, he began to view all education, not just science, as a means of creating moral, virtuous citizens.

Public Life

The Herschels returned to England in 1838, by which time John was in his mid-forties. His focus upon his return was helping his children prepare for adult life. He took on government roles too, but most immediately he became involved in a brand-new, cutting-edge technology: photography. Long before getting married and travelling to South Africa, John's chemistry work had led him to a discovery of the chemical 'hypo' or 'hyposulphate of soda'.[52] Back in England, he now found a use for it for 'fixing' photographs and suggested this process to his friend Henry Fox Talbot (1800–77) who adopted it from the 1840s onwards (see Chapter 7).

The Herschels did not return to Observatory House in Slough; instead, they moved to a larger house called Collingwood in Kent to accommodate their growing family. Maria Sophia, Amelia, Julia, Matilda, and Francisca were all born between 1839 and 1846, bringing the total number of Herschel children to eleven. Almost a decade later, in 1855, they were joined by Constance. All the children were taught at home using a combination of nurses, governesses, tutors, and parents. How these lessons were divided depended somewhat on the expertise of the particular governesses employed at any time, but as a general rule John taught advanced mathematics and Latin, showed the children experiments, and took them to scientific lectures. As the boys got older, he arranged future careers for them and the schooling that would lead them there. William, thanks to Margaret's guardian Eneas Mackintosh, joined the Indian Civil Service; Alexander it was hoped would make a career of science; and John Junior combined the two, eventually going to India as an engineer.

Outside of family life, John continued the role he had begun in South Africa as a government advisor on education. He became Master of the Mint in 1850 and from 1850 to 1851 participated in the Cambridge Commission set up to examine teaching at Cambridge.[53] John was instrumental in the Commission's conclusions, which included the recommendation of introducing of a new set of exams in natural sciences and moral sciences. (Previously all students had only been examined in the Mathematics Tripos.) It is unclear why John took up the post once held by Sir Isaac Newton as Master of the Mint. Conversations with his friend Richard Jones suggest that with his observations from the Cape finally published (a task that took him almost ten years to complete), he wanted a change. He sought a role involving some kind of public service, and Jones suggested Master of the Mint.[54]

This civil service position put him in charge of staff at the Treasury and ultimately made him responsible for the currency of the entire empire. Unfortunately, his primary role was to push through unpopular restructuring among the workforce. John was not used to this kind of bureaucratic responsibility and did not enjoy it. It was also a post that took him away from his family and his scientific work, as he had to live in London away from the family home of Kent. His health suffered because of the stress, and by 1855 he stepped down. Returning to family life, he delighted in the time he was able to devote to his youngest daughter's education, which he continued to be involved with until his death in 1871. He died aged seventy-nine at home, surrounded by his family and given a national funeral and burial in Westminster Abbey. When he died, his youngest daughter Constance was sixteen years old. Her mother wondered how to complete an education that had had such a promising start. In the end, she decided to send her to the newly opened women's college, Girton, at Cambridge University, and so John's youngest daughter became one of the first women to study at Cambridge.[55]

Conclusion

John Herschel never fit neatly into the ideal of a heroic man of science. He made no single great discovery on which to hang his name. Nor did he excel in a single field to the exclusion of others. But this is what makes him such an interesting character to study. His background and upbringing meant he could have focused exclusively on astronomy or mathematics, carrying on the work of his father and aunt, but he chose to go beyond this. He advanced mathematics in new ways, then questioned the purpose of what he was pursuing and moved on. He tried chemistry and mineralogy but then chose family and the chance to work with his father while he still could. He looked beyond his father's projects in astronomy to see what else he could do with it and how he could use his position and influence to better society by making science more central to it. Then, when he married, he allowed family life to take up significant portions of his time and energy and so give less of his time and attention to new projects. Some have seen Herschel as the last of the natural philosophers, rich enough to have sufficient time and energy to dabble in a broad range of scientific areas. If he was, then ironically he helped close that door behind him, creating a scientific world that valued those with access to universities and professional opportunities over those operating in a more domestic sphere.

In some ways, there is a gendered element to Herschel's story. We are used to talking about women balancing family and professional life, but surely this must have been true of some men too. John Herschel's story, especially the way in which it appears to change after marriage, certainly suggests this was true for him. John grew up in an environment in which science and domesticity were intertwined. His father and aunt lived and worked in a home that was also observatory and workshop. Herschel, in contrast, grew into – and indeed was instrumental in creating – a world in which science was separate from the domestic sphere and instead took place in professional, public spaces.

John Herschel's story also shows what a life in science looked like in the nineteenth century, what options were available and how they changed over time (changes in substantial part due to his campaigning). He tried and failed to earn a living in science, and then he helped others by changing how science was viewed and valued through his writing, through his involvement in public bodies such as the Astronomical Society and the Cambridge Commission, and through his public persona. Although not as well-known today, in his day he typified science for the public and for that reason alone makes an ideal avenue through which to better understand nineteenth-century British science.

Notes

1 M. J. Crowe, ed., *A Calendar of the Correspondence of Sir John Herschel* (Cambridge: Cambridge University Press, 1998), 1. For biographical information, see Emily Jane Winterburn, 'The Herschels: A Scientific Family in Training' (PhD diss., Imperial College London, 2011).

2 Susan [Walter F.] Cannon, 'John Herschel and the Idea of Science', *Journal of the History of Ideas*, 22 (1961), 215–39.

3 Caroline L. Herschel, 'Biographical Memorandums of My Nephew J. Herschel to the Best of My Recollections', 27 May 1838, British Library (BL): microfilm M/588(4).

4 Herschel, 'Biographical Memorandums,' BL: microfilm M/588(4). 'Shew me the Wail!' likely refers to the illustration of Cetus in Flamsteed's atlas of constellations.

5 Caroline L. Herschel to Margaret Brodie Herschel, 6 September 1833, BL: Eg3761 f189–90.

6 J. R. Oldfield, 'Private Schools and Academies in Eighteenth-Century Hampshire', *Proceedings of the Hampshire Field Club Archaeological Society*, 45 (1989), 147–56.

7 Herschel, 'Biographical Memorandum'.

8 John Herschel to Mary Pitt Herschel, 9 August 1799, in 'Newbury' envelope, John Herschel-Shorland (JHS) papers, Box Files Right 4.

9 Painting in the Herschel-Shorland family's private collection, JHS, objects in Games Room.

10 Eton College Register, 1753–90, p. 422. Internet Archive, https://archive
    .org/details/etoncollegeregis00austuoft/page/422/mode/2up, accessed 28
    January 2023.
11 Family tree from the Herschel-Shorland family's private collection.
12 Martha McMackin Garland, *Cambridge before Darwin: The Ideal of a
    Liberal Education, 1800–1860* (Cambridge: Cambridge University
    Press, 1980).
13 Caroline L. Herschel to Margaret Brodie Herschel, 2 June 1841, BL:
    Eg3762 f78–79.
14 John Venn and John Archibald Ven, *Alumni Cantabrigienses* (Cambridge:
    Cambridge University Press, 1922–54).
15 Caroline L. Herschel to Margaret Brodie Herschel, 7 July 1842, BL:
    Eg3762 f104–5.
16 All that is known of Dr Hausann is an entry in the Herschel accounts stating
    a payment to him was made for 'teaching my son German'. Slough House
    Book, National Maritime Museum: HRS/2–3.
17 Mary Pitt Herschel's travel journal 1809 (entry for 2 August), JHS
    Papers, Armoire.
18 John Herschel's travel diary 1809–10, HRC, Herschel Family Papers,
    Container 22.1.
19 Mary Pitt Herschel's travel journal 1809 (entry for 22 August), JHS
    Papers, Armoire.
20 Mary Pitt Herschel's travel journal 1809 (entry for 23 July), JHS
    Papers, Armoire.
21 John Herschel to John Whittaker, 2 July 1813, St John's College Library, 4.
22 John Herschel to Emilia Stewart, 4 December 1828, JHS Papers, Box
    Files Right.
23 Caroline's diary entry for 17 October 1809, quoted in Constance Lubbock,
    *The Herschel Chronicles: The Life Story of William Herschel and His Sister
    Caroline Hershel* (Cambridge: Cambridge University Press, 1833), 324.
24 Observatory House Visitors Book, 1783–1828 (copied), JHS Papers, Box
    Files Left.
25 Alexander Rogers to John Herschel, 5 March 1810, RS:HS 14.
26 John Herschel to William Nicholson, 23 March 1812, published in the
    Journal of Natural Philosophy, Chemistry and the Arts, 32 (1812), 13–16.
27 Andrew Warwick, *Masters of Theory: Cambridge and the Rise of
    Mathematical Physics* (Chicago: University of Chicago Press, 2003), 77.
28 P. C. Enros, 'The Analytical Society (1812–1813): Precursor of the Renewal
    of Cambridge Mathematics', *Historica Mathematica*, 10 (1983), 24–47,
    quote on 27.
29 Laura J Snyder, *The Philosophical Breakfast Club* (New York: Broadway
    Books, 2011).
30 John Herschel to Charles Babbage, 14 July 1816, RS:HS 2.64.

31 Jean-Baptiste Biot to John Herschel, 2 May 1819, RS:HS 4.84. John Herschel to David Brewster, 15 May 1819, RS:HS 20.70.

32 William J Ashworth, 'John Herschel, George Airy, and the Roaming Eye of the State', *History of Science*, 36 (1998), 151–78, at 163.

33 James Secord, *Visions of Science: Books and Readers at the Dawn of the Victorian Age* (Chicago: University of Chicago Press, 2014), 102.

34 John Herschel, *A Preliminary Discourse on the Study of Natural Philosophy* (Chicago: University of Chicago Press, 1987 [1830]), 22, 25.

35 Janet Browne, *Charles Darwin: The Power of Place* (Princeton: Princeton University Press, 2003), 180.

36 M. Jeanne Peterson, *Family, Love and Work in the Lives of Victorian Gentlewomen* (Bloomington: Indiana University Press, 1989); Pnina G. Abir-Am and Dorinda Outram, *Uneasy Careers and Intimate Lives: Women in Science 1789-1979* (New Brunswick: Rutgers University Press, 1987).

37 Margaret's father, Alexander Stewart, was married twice, first to Louisa Macpherson then to Emilia Calder. From the first marriage, they had two children: Alexander and Catherine. From the second marriage, they had seven children: Charles Calder, Duncan, James Calder, Patrick, Margaret Brodie, Isabel, and John.

38 Charles Stewart to Margaret, 9 February 1827, JHS papers BFR[?].

39 British Library, India Office Papers, L/MIL/9/376 f.123; N/1/39 f.125 'Letters from China 1835-36', *Journal of the Hong Kong Branch of the Royal Asiatic Society*, 11 (1971), 52–61.

40 Charles Stewart to Emilia Stewart, 16 May 1827, JHS papers, Box files right. John Bird, *The Juvenile Observatory* (Eton: E. Williams, 1824).

41 Sara Delmont, *Knowledgeable Women: Structuralism and the Reproduction of Elites* (London: Routledge, 1989), 2.

42 Augustus J. C. Hare, ed., *The Life and Letters of Maria Edgeworth* (London: E. Arnold, 1894), 179–80. Mrs R. Butler to Harriet Edgeworth, 29 March 1831.

43 John Herschel to Emilia Stewart, 4 December 1828, JHS papers, Box files right.

44 John Herschel to Mr Roberts, undated [1830?] Yale University. John Herschel to Margaret Herschel, August 1830, BL: microfilm M/588 (1).

45 John Herschel to James Calder Stewart, 11 January 1831, HRC: folder Lo412.

46 W. T. Ferguson and R. F. M. Immelman, *Sir John Herschel and Education at the Cape, 1834-1840* (Cape Town: Oxford University Press, 1961), 41, quoting John Herschel to Rev Dr J. Adamson, 21 November 1835.

47 Zachary MacAulay, 'Natives of South Africa', *Anti-Slavery Monthly Reporter*, 50 (July 1829), 30–31.

48 Margaret Herschel to Emilia Stewart, 26 September 1834, from Brian Warner, ed., *Lady Herschel Letters from the Cape, 1834-1838*, (Cape Town: Friends of the South African Library, 1991), 49–50.

49 Margaret Herschel to Emilia Stewart, 11 July 1834, in Warner, *Lady Herschel*, 42. 'Hottentots' was a colonial name for the indigenous people of the Western Cape, then called the Khoikhoi. Margaret Herschel to Emilia Stewart, 11 April 1835, and Margaret Herschel to Duncan Stewart, 6 June 1835, Warner, *Lady Herschel*, 69, 76.

50 Despite spending many hours at night surveying the sky and many more in the day helping to educate their children, John and Margaret still found time to carry out an additional project together. They studied the local plants and flowers, making careful, beautiful drawings of each specimen. The hundreds of drawings they made were never published in their lifetime but have since been put together in a volume published alongside highlights from their prints and drawings collection. Brian Warner, ed., *Flora Herscheliana* (Johannesburg: Brenthurst Press, 1998).

51 John Herschel to Charles Davidson Bell, 17 February 1838, in Warner, *Lady Herschel*, 6.

52 Syndey Ross, 'Herschel and Hypo', *Nineteenth-Century Attitudes: Men of Science* (Alphen aan den Rijn: Kluwer, 1991), 173–99.

53 'Papers relating to the Cambridge Commission', BL: microfilm M/588 (7).

54 Snyder, *Philosophical Breakfast Club*, 299.

55 Letters written by Margaret Brodie Herschel held with Herschel family, JHS paper, Box files right.

2

The Mathematical Journey of John Herschel

John Herschel's destiny lay in science. Ahead was a career of creation in astronomy, botany, chemistry, geography, meteorology, photography, and much else. Not least would be the attention he paid to mathematics. His royal road to becoming a leader in Victorian science was paved by William and Caroline Herschel, his celebrated father and aunt, both of whom were acknowledged authorities in astronomy. William, a German émigré to England, held the position of 'King's Astronomer' and in this role established a rapport with George III, an amateur astronomer himself. William gained wider recognition as the discoverer of 'The Georgian' (known later as the planet Uranus) and Caroline as the discoverer of comets.

When John was born in 1792, the Herschel family lived in Observatory House on the Windsor Road in Slough, on whose grounds his father had built a giant forty-foot reflecting telescope, a local landmark completed in 1789. Slough was a quiet place in pre-Betjeman days when stagecoaches passed along the Great West Road linking London with the West of England. As the King's Astronomer, it was important for William to live near Windsor Castle, where George III resided.[1] Before astronomy, William had taken to mathematics, and it soon became an obsession, overtaking his professional life as a musician. He added mathematical books to his library, so as John was growing up there was an ample supply of the classics of mathematics at hand.

John Herschel's formal schooling was disjointed, divided between different schools and private tutors (see Chapter 1). When these were unable to furnish a suitable level of mathematics, Scotsman Alexander Rogers was employed in 1806 as a private tutor to teach the boy mathematics and French. An accountant by profession, Rogers was an able mathematician and astronomer. It is likely he shared the same opinions as his patron in Scotland, John Playfair, widely known for his attacks on the poor state of English mathematics.[2] Rogers urged his fourteen-year-old pupil to read Joseph-Louis Lagrange's *Mécanique analytique* in preparation for the study of Pierre-Simon Laplace's *Mécanique céleste*, a benchmark in the mathematical treatment of astronomy.[3] In so doing, he opened the boy's eyes to modern developments in the science of the cosmos and how it could be treated mathematically.

With their European background, the family was well placed to appreciate scientific developments being made outside England. In July 1802, they visited Paris, where William met leading scientists including Laplace. In Paris, he had an audience with Napoleon, where he and the First Consul exchanged scientific pleasantries on astronomy at his palace at Malmaison. Through these multi-layered connections, John saw at first hand the importance of community and the exchange of ideas fuelling scientific progress. In addition, to show him the advances mechanisation was having on industry in the emerging factory system in Britain, his father took him on a trip to the north of England. This wide-ranging educational experience helped John Herschel combine his obvious abilities in mathematics with a practical side, and this manifested itself in a balance between his attention to mathematical theory, the practical needs of observational astronomy, and such emerging technologies as photography.

Undergraduate Life

Entering St John's College in October 1809, John Herschel was well ahead of the normal freshman in terms of mathematical expertise. He arrived with a pronounced ability in the subject and how it might be applied to the advancement of astronomy. He had been introduced to 'French analytics' by Rogers, whereby the study of mechanics was carried out using rigorous calculus methods, such as analysing solutions of differential equations, as opposed to geometrical methods prevalent at Cambridge. Herschel soon realised that the Cambridge Mathematical Tripos course of mathematics, steeped in its eighteenth-century geometric tradition, would be of little help in understanding modern ideas leading on from the material Rogers had taught him.

The business of Cambridge University was not the education of future mathematicians but the training of clerics and the professions. In the main, mathematics was seen by examination-oriented dons as a static body of knowledge laid down by Newton and the ancient Greeks and one that logically unfolded from first principles. There was little recognition of mathematics as a developing subject or appreciation of mathematical advances taking place on the Continent. There were some reform rumblings, notably by Robert Woodhouse (1773–1827), but this was a limited response and isolated.[4] One Cambridge tutor, John Toplis (c.1775–1857), had translated the first part of the *Mécanique céleste* in 1814, but again this had little impact on the teaching of the mathematical curriculum.[5]

Amongst the teaching dons in general, there was scarce appreciation of modern French analytics. The Mathematical Tripos course, culminating in the Senate-house Examination, was the oldest course of study in the university and had tradition on its side; it had been shaped during the previous century and any change of it was fiercely resisted. Herschel's assessment of the exact sciences in England as it existed at the end of the eighteenth century and the beginning of the nineteenth was damning: 'Mathematics were at the last gasp, and Astronomy nearly so.' Herschel was entering a Cambridge that offered little more than the 'chilling torpor of [teaching] routine'.[6]

The textbook staples used at the end of the eighteenth century reflected this environment: Euclid's *Elements* was highly valued and Newton's *Principia* revered. Newton had viewed the calculus from its use in dynamics and, for example, wrote velocity as the 'fluxion' $\dot{x}$ of the distance x. Thus $\dot{x}$ is the derivative, where it is understood that the 'fluent' x (distance) flows through time. Following Newton, fluxions were taught exclusively with this 'dot-age' notation, with reliance placed on intuition gained from geometric figures. As Herschel had signalled, mathematics teaching at Cambridge at the start of the new century was in the doldrums.

It was against this background that textbooks were written for the use of undergraduates. Samuel Vince (senior wrangler 1775,

1749–1821) produced *A Treatise on Practical Astronomy* (1790), *Principles of Fluxions* (1795), and *Principles of Hydrostatics* (1796), all of which used the synthetic geometric style as opposed to the new analytical perspective adopted on the Continent. Apart from their 'want of elegance', these textbooks could not cope with the modern analytical theories of their target areas of application. In a similar vein, James Wood (senior wrangler 1782, 1760–1839) published *Principles of Mechanics* (1796) and *Elements of Optics* (1798), though his *Elements of Algebra* (1795) at least gained praise for its clarity.[7]

Herschel was justly scathing about the content of the Mathematical Tripos and called for 'the termination of this childish course of study'.[8] He saw mathematics as a subject to be researched, not simply memorised in order to pass examinations This commitment manifested itself in two publications he made as an undergraduate. In February and May of his third year at the university (1812), he submitted letters to the editor of Nicholson's *Philosophical Journal* and signed them as 'A Lover of the Modern Analysis' and 'Analyticus'.[9] Published anonymously, these short papers dealt with infinite series expansions and infinite products of various trigonometric functions, from which he deduced mathematical identities originally stated by John Wallis (1616–1703) and Leonhard Euler (1707–83). Herschel said the works had afforded him 'some degree of amusement' but played down their importance as 'trifles'.

The Analytical Society

In this context, the formation of the undergraduate Analytical Society was a beacon for reform. Its initial meeting took place at Gonville and Caius College on 7 May 1812.[10] Its exact composition is unclear, but it is probable that membership never exceeded a dozen undergraduates at any one time. However, the Society attracted proficient students who went on to leading positions in the academic world and would influence the next generation. The Society's main drivers were Herschel with Charles Babbage (1791–1871) and George Peacock (1791–1858) alongside. Babbage and Peacock were from Trinity College and Herschel from St John's; Babbage and Herschel were twenty years old and Peacock twenty-one.

At the inaugural meeting, Herschel, already regarded as the potential senior wrangler of 1813, was elected president. Enthusiasm for Continental mathematics motivated the group, and, as its name indicates, the objective of the Society was the promotion of the analytical style of mathematics, symbolic and algebraic in nature as opposed to the existing Cambridge synthetic style based on geometry. Discussions at Society gatherings were lively, and members were imbued with a strong sense of missionary zeal. In correspondence with Babbage, Frederick Maule (1790–1813), one of the founders of the Society, remembered meetings and the 'boisterous debate so well described in your letter that I transported myself in thought to the scene of action and heard the damns, the nonsense, the arguments, the objections'.[11]

The experiences of student Alexander d'Arblay (1794–1837) provides another sidelight on the Society. In October 1813, the eighteen-year-old only child of novelist Fanny Burney and Parisian exile General Alexandre d'Arblay was admitted to Gonville and Caius College, his mother bringing him to England to avoid conscription in the French army. At Cambridge, a mathematical culture shock was in store, for at school in Paris he had been taught mathematics through analytics. His aunt wrote to his mother that Alexander should be persuaded 'to study in the Cambridge way, that is to say, to learn to solve his problems and to give their proofs by geometry instead of algebra or the analytical method, which is the French way and also the best; and Alex knows that'.[12] These sympathies help expose the objectives of the Analytical Society, and though it remains unclear whether d'Arblay formally joined the Society, he was a close associate.

By October 1812, Herschel was beginning the 'tenth term' of the Cambridge course, the period when competitive students were preparing themselves for the Senate-house degree examination held in January. Despite the looming January deadline, Herschel read a paper to the Analytical Society on Cotes's theorem, an analytical result based on the circle, which he extended to conics and in particular the parabola.[13] The theorem had been discovered by Roger Cotes (1682–1716) a hundred years previously and recently proved by John Brinkley (1766–1835).[14] In Herschel's hands, Cotes's theorem grew into a substantial paper communicated to the Royal Society by his father and printed in the *Philosophical Transactions* of 1813.[15] That year Herschel enjoyed a somewhat meteoritic rise in the English scientific world, gaining celebrity status on account of being the year's senior wrangler (top student) at Cambridge along with election to fellowship of the Royal Society of London. William Whewell (1794–1866), then a second-year student at Trinity College who met him that year, wrote home that Herschel was 'a most profound mathematician and an excellent general scholar'.[16]

In addition to Society meetings, there was a clear ambition among its membership to set the Society's views and record in print and thus form a definite challenge to Cambridge's geometric orthodoxy. Disputation at meetings was one thing but writing was another. Despite the collective enthusiasm from members, Herschel and Babbage found themselves writing what turned out to be the single volume of the *Memoirs of the Analytical Society*. Proposed in November 1812, the *Memoirs* was forged over the summer of 1813 with Herschel and Babbage in constant communication with each other. There were doubts along the way, and at times Herschel suffered despondency. At the outset, he wrote to fellow Society member Edward Ffrench Bromhead (1789–1855) of his misgivings about the whole project:

> I am by no means so sanguine, although not less sincerely desirous of contributing to the introduction of a better taste in analytics than at present prevails. – The ill success of a first undertaking [of] (the Anal. Soc.) although it has not in the least damped my ardour in this respect, has yet a good deal sobered it. – The fire of enthusiasm spreads only where it meets with inflammable matter to receive & cherish it – and how few, how very few are those who are disposed to enter heart & soul into a task of such gigantic labour, and such diminutive reward. Of that few again, how small a proportion have the time or the peculiar turn of mind so necessary to realise their plans . . .. The publication of a Mathe [matica]l work, particularly if it goes one step beyond the comprehension of Elementary readers is a dead weight & a loss to its author.[17]

Despite the difficulties, the *Memoirs* was published a year later: over one hundred pages of dense mathematics ready for wider readership. A 'Preface' describing the merits of analytics (and a brief history) was jointly written by Herschel and Babbage. Its opening explains the overarching philosophy of the Society and what it was attempting:

> To examine the varied relations of necessary truth, and to trace through
> its successive developments, the simple principle to its ultimate result, is
> the peculiar province of Mathematical Analysis. Aided by that refined
> system, which in the ingenuity of modern calculators has elicited, and to
> which the term Analytics is now almost exclusively appropriated, it
> pursues trains of reasoning, which, from their length and intricacy,
> would resist for ever the unassisted efforts of human sagacity.

The articles in the *Memoirs* appeared anonymously to camouflage the
fact that the journal was written by just two people.[18] With one eye on
the turbulent political situation of the time and the war with France,
Herschel expressed their role as 'the ringleaders, if not the only actors in
this literary assault upon the peace and quietness of the world'.[19] For
both he and Babbage, the superiority of analytics as a methodology lay
in the 'accurate simplicity of its language' as an instrument of reason.
Following the Preface, Herschel supplied papers in the *Memoirs* 'On
Trigonometric Series' and 'On Equations of Differences'. Exploration of
the calculus of differences featured in much of his early writing. This
was distinct from the differential calculus though closely connected
with it. The thrust of the calculus of differences, the principal accom-
plishment of Herschel's work in pure mathematics, is based on the
difference operator Δ_h, defined by $\Delta_h f(x) = f(x+h) - f(x)$ and its
variants. Operators could be treated as entities in themselves, with a
distinction between the operator and the quantity on which it operates.

Though the zealousness of Analytical Society members was apparent, mathematics as 'analytics' and the ensuing technical developments
proved too difficult for most readers. Bromhead warned Babbage in
1813 that 'not one mathematician in 10^∞ can understand [analytics]'.[20]
Bromhead found further evidence of this during a later trip to
Cambridge in 1816, reporting back to Babbage, 'I did not find a single
soul, even among Senior Wranglers, Herschel excepted, who understood a word about it'.[21] This difficulty proved the undoing of the
Memoirs, and for this reason alone it is not surprising that it received
poor recognition.[22] Its mathematical contents had no connection with
the Mathematical Tripos at Cambridge, and no teaching don would
recommend it to their students. In addition, it was a financial flop,
priced at the unrealistically high amount of fifteen shillings per copy.
As Whewell summarised:

In this publication, the extraordinary complexity and symmetry of the symbolical combinations sorely puzzled the yet undisciplined compositors of that day and led unmathematical readers to the conviction that the whole was a wanton combination of signs, left to find a meaning for themselves, like the Javanese character of Princess Caraboo.[23]

The Analytical Society did not survive. After graduation, members went on their way to such careers as the church, the law, or, as was the case with Bromhead, back to their landed estates. Herschel wanted to move the Society to London, but this goal was not realised. In terms of a career, he found mathematics was not the best way forward, and this contributed to his decision that the life of a Cambridge don was not for him. His father advised a clerical career, but this did not sit well with him, and instead he opted for legal training. In January 1814, he entered Lincoln's Inn in London to train as a Chancery barrister.[24] In the metropolis, he mixed in scientific circles and was influenced by the chemist William Hyde Wollaston (1766–1828) and the astronomer-physician James South (1785–1867). These influences broadened his scientific outlook and in the long term diminished his focus on mathematics.

At the time he joined the legal profession Herschel published 'Consideration of Various Points of Analysis', a paper on the calculus of generating functions (*fonctions generatrices*).[25] In it he cited Laplace and Gaspard Monge (1746–1818) as well as the Scottish mathematicians Thomas Knight (1775–1853) and William Spence (1777–1815). The revolutionary part of this subject was expressed by his use of the functional notation 'together with the method of separation (where it could conveniently be done) the symbols of operation from those of quantity' and with the 'grand and ultimate object, the union of extreme generality with conciseness of expression'.[26] It was the first paper Herschel wrote as a fully-fledged member of the Royal Society, read to the Society in May 1814.

At Lincoln's Inn, Herschel's appreciation 'of the dry details of law' caused a change of heart in the summer of 1815 and the abandonment of the legal profession. He returned to Cambridge, where he was appointed as a lowly sub-tutor in mathematics at St John's College. But he soon tired of this, writing to Babbage of 'examining 60 or 70 blockheads ... not one in ten of whom knows anything but what is in the book ... and I have not made one of my cubs understand what I would have them drive at'.[27] During this time, he produced his second paper as a member of the Royal Society, on exponential functions, read in December 1815.[28] This paper contains 'Herschel's theorem', a theorem giving the expansion of $f(e^x)$ as a power series. Noted by later mathematicians such as Augustus De Morgan (1806–71), William Rowan Hamilton (1805–65) and George Boole (1815–64) as useful in the calculus of functions and the differential calculus, it became well known in the nineteenth century but has now almost disappeared from view.[29]

Leaving Cambridge

Teaching was not to Herschel's taste, and in September 1816 he wrote to Babbage, '[I am] going under my father's direction to take up star gazing.'[30] As the prospect of another academic year loomed, he returned to the university to say goodbye: 'I shall go to Cambridge on Monday where I mean to stay just enough time to pay my bills, pack up my books and bid a long – perhaps a last farewell to the University ... I always used to abuse Cambridge ... but upon my soul, now I am about to leave it, my heart dies within me.'[31] He duly took up astronomical observations directed by his father, who, in his seventies and declining in health, was unable to put in long hours at his telescope in all weathers. Yet even during this apprenticeship Herschel still made time for mathematics. In November 1817, he completed a paper on the calculus of differences, 'On Circulating Functions', a subject he would return to in his mature years.[32] In this paper, he reiterated the power of analysis, 'the universal medium of mathematical enquiry'.[33]

Though the single volume *Memoirs* of the Analytical Society made little impression, the next major publication in the cause of analysis produced by the Herschel-Babbage-Peacock trio was a success and met the educational need of serious Cambridge students. Writing in Nicholson's *Philosophical Journal* in 1812, an enquirer had been curious about the new 'modern analysis'. The student had been grounded in pure geometry and the calculus built on Newton's 'fluents and fluxions' and was unaware of mathematical developments on the continent of Europe. He was duly advised to consult *Traité élementaire du calcul differentiel et du calcul intégral* (first edition 1802, second edition 1806) written by Sylvestre François Lacroix (1765–1843), but the only option at the time was to study the book in its original French.[34] Clearly, there was a need for a comprehensive textbook treatment of this new mathematical approach in English.

In late 1815, Babbage, Peacock, and Herschel contemplated satisfying this need by producing a translation of Lacroix's *Traité élementaire* customised for the Cambridge course of study. They hoped it would be useful to 'men cramming for degrees' and that the resulting *Treatise* would hold its own in the competitive market of Cambridge textbooks – and perhaps even replace some of the geometrically oriented textbooks in current use. In addition, Herschel would supply an extensive appendix on finite differences based on the *Traité* and his own previous work.[35]

In the original full three-volume *Traité* on which the *Traité élementaire* was based, Lacroix had adopted the Lagrangian basis for the calculus, in which the notion of a mathematical limit was avoided. In this approach, successive derivatives of a function were defined as coefficients in the Taylor expansion of the function. By this means, calculus without the limit concept became a branch of algebra. When Lacroix came to write the *Traité élementaire*, he switched to the 'limit-based approach' in which the theory of limits was used as a basis. In their textbook translation, Herschel and his collaborators switched back to the algebraic version promoted by Lagrange. The Analytical Society members saw this as providing the proper logical basis for the calculus, but in doing so, they adopted a system of little use for the purposes of 'mixed mathematics' where calculus was to be applied to physical problems.

Unsurprisingly, introduction of the translation at Cambridge was met with opposition. Trinity College Fellow Daniel Mitford Peacock (c.1768–1840) was a conspicuous opponent of the introduction of the translated *Treatise* into the Cambridge curriculum and criticised it for its abstractness, warning against its use in the Mathematical Tripos. 'Academical education', he wrote, 'should be strictly confined to subjects of real utility'. The eminent natural philosopher Thomas Young (1773–1829) was another voice resisting the incoming 'Analytics'.[36] One Cambridge man recalled his reactions to the introduction of the 'new math'. James Challis (1803–82), future Cambridge professor of astronomy, remembered his experience as a freshman:

> When I commenced the study of the Differential Calculus, the introduction of the analytical method of reasoning was, as I well remember, still regarded with jealousy by the older mathematicians of the University, from the apprehension that it was not susceptible of the rigid reasoning they considered to be characteristic of the geometrical method, especially as exemplified in Newton's *Principia*.[37]

Despite this hostility, the publication of the *Treatise* was so successful that the three translators considered providing supplements. They produced *Collections* of examples of the applications of analysis in their chosen areas: Herschel on finite differences, Peacock on calculus, and Babbage on functional equations. Peacock playfully reminded Herschel there might even be hopes of financial reward: 'I think we may fairly calculate that this work will constitute a little annuity to us, of at least 25 £ per annum for each of us: a very considerable sum & and when we consider the present state of the country, it will be as good as a farm which pays us rent.'[38]

Of the three authors, Peacock remained at Cambridge and was closest to the development of the undergraduate curriculum. He was moderator (setter of examination questions) of the Mathematical Tripos on three occasions. At the first examination, in 1817, he made the bold move of introducing Continental notation of the differential calculus into the examination, much to the consternation of the more traditionalist dons. This notation was quickly adopted, and within two years the 'antiquated fluxional notation' had disappeared from the examination. By the beginning of the 1820s, the analytic revolution had significantly progressed in Cambridge.

For his work in mathematics, Herschel was awarded the Copley Medal, the Royal Society's highest honour, in 1821. He was twenty-nine years old, and the award capped the mathematical work of his youth. Though he occasionally returned to mathematics in subsequent years, he had completed his most creative mathematical work and looked to concentrating his energies elsewhere.[39] In scientific work, he turned to astronomy, chemistry, optics, and the study of crystals (see Chapters 3, 7, and 8).

Looking Back

Though Herschel moved on from mathematics, in the 1820s he continued to be thought of as a mathematician. When the Lucasian chair of mathematics at Cambridge became vacant in 1826, he was urged to consider it. Though Whewell thought the stipend 'rather a starving matter', the holder would enjoy the prestige of occupying a chair once held by Newton. In the end, neither Herschel nor Whewell was appointed. Whewell's tutor position prevented his application, whereas Herschel's absorption with astronomy and the cataloguing of double stars ruled him out.[40]

Fifteen years after leaving Cambridge, Herschel remembered his previous mathematical life and his thought on mathematics. In an encyclopaedia article on the subject, he elaborated the 'two great branches' of mathematics. The first, 'demanding no assistance from inductive observation, and very little from the evidence of our senses, constitutes the *pure mathematics*'. The other, 'taking for granted the truth of general laws deduced by legitimate induction from observations sufficiently numerous, supplies the hidden links which connect the cause with its remote effect, and endeavours from the intensity of the one to estimate the magnitude of the other. To this branch of mathematics the epithet *mixed* is in consequence applied'.[41] Herschel's early contribution was to pure mathematics, but his general skills and confidence in handling intricate numerical data was brought to bear in the second branch: applying those techniques to astronomy.

Upon his return from the Cape of Good Hope in 1838 and elevated to the rank of baronet in the British social system, Herschel became emblematic of what it meant to be 'scientific' in the opening years of Victorian England (see Chapter 4). He was showered with honours and invitations to membership of scientific societies at home and abroad. When the British Association for the Advancement of Science (BAAS) held its annual meeting in Cambridge in 1845, Herschel was elected the meeting's president. In his address, he touched on mathematics, noting the new discoveries being made in algebra, such as the quaternions and octaves, that resulted in 'conceptions of a novel and refined kind ... introduced into analysis'. He cited De Morgan and Hamilton and now praised Cambridge as – far from being a mathematical backwater – the place where further developments in mathematics would naturally take place.

Herschel saw developments at Cambridge as the continuance of his own attempts to reform mathematics. 'I look back to the vast and extraordinary development in the state of mathematical cultivation and power in this University', he told the gathered BAAS members, 'as evidenced both in its examination and in the published works of its members, now, as compared with what it was in my own time'. One evidence of the change that had taken place was the founding of the *Cambridge Mathematical Journal* in 1837, and Herschel in particular highlighted the papers of its co-founder Duncan F. Gregory (1813–44) on the calculus of operations.[42]

By now Herschel was a scientific statesman, having made contributions to many branches of science and by virtue of his position within the British establishment. He maintained contact with Cambridge and served as one of the Royal Commissioners looking into the state of the university (see Chapter 10). He contributed a range of essays to publications such as the *Edinburgh Review* and the *Quarterly Review*. One of these, published in 1850, was his lengthy review of 'Theories of Probability' by Adolphe Quetelet (1796–1874). In this essay, Herschel discussed the distribution of error, the relevance of the normal distribution of probabilities, and the circumstances under which they can be multiplied. These considerations surfaced again when James Clerk Maxwell (1831–79) was formulating his kinetic theory of gases in 1859, possibly due to the influence of Herschel's essay.[43]

As his other scientific work progressed, Herschel did not lose sight of mainstream mathematics. He continued to take an active part in it, especially when it was realised that his early work was relevant to the study of partition theory, a field of increasing interest to English mathematicians in the middle years of the nineteenth century. Partition theory has a long history and included contributions by such mathematical leaders as Euler and Leibniz prior to Herschel and G. H. Hardy (1877–1947) and Srinivasa Ramanujan (1887–1920) after. The basic problem is of counting the number of ways a positive integer n can be written as the sum of positive integers. For example, for $n = 7$ one way is $7 = 1 + 2 + 4$ and another is $7 = 1 + 3 + 3$. In the first case, 7 is written in terms of distinct parts, whereas in the second 7 is written as the sum of odd parts. There are many significant theorems in partition theory, such as that the number of partitions of n into distinct parts is equal to the number of partitions of n into odd parts. In Herschel's time, prominent English contributors to this field included De Morgan, Henry Warburton (1784–1858), Norman MacLeod Ferrers (1829–1903), and especially Arthur Cayley (1821–95) and James Joseph Sylvester (1814–97). In 1851, Herschel delivered a paper to the Royal Society returning to an earlier result of his of finding a periodic term in the counting function $p(n)$ of partitions of n. Though his method is laborious, it anticipated Cayley's more refined treatment of the same subject in which Cayley introduced 'prime circulators'.[44] Herschel returned to the topic of finite differences with a 1860 paper in the *Philosophical Transactions*, to which Cayley added a note bringing it up to date by alluding to work by Gotthold Eisenstein (1823–52) and Sylvester.[45] Whereas Cayley and Sylvester worked intensively (and competitively) on the theory of partitions during this period, Herschel had anticipated them both in providing formulae for $p_k(n)$, the number of partitions of n into at most k parts in the case of small values of k.[46]

Epilogue

As second president of the newly founded London Mathematical Society (1865), Sylvester wrote to Herschel expressing the wish that he join the Society: 'It would be a source of universal gratification to its members if you should feel disposed to add your honored name to their number, and it is thought that you would thereby be doing good service to the cause of Mathematical Studies in this Country.' Sylvester would have remembered that Herschel was one of his sponsors on his election to fellowship of the Royal Society in 1839. He also had first-hand knowledge of Herschel's mathematical work. Herschel would certainly have lent the Society added prestige on account of his place in the scientific world. Unfortunately, it is almost certain he did not join even though he had not given up mathematics.[47]

Given Herschel's history with analysis, it is curious that geometry was the subject of one of his final forays into mathematics. In 1859, Herschel approached the Geographical Society with a new projection of the sphere based on a partial differential equation of the second order. He claimed it was superior to the tried and tested Mercator projection widely used in cartography. Though he quoted Gauss's work in connection with a Prize Question proposed by the Royal Society of Copenhagen in 1822, he seemed oblivious of a similar projection made by Johann Heinrich Lambert (1728–77) in 1772.[48]

Herschel's attention was drawn to a myriad of subjects, but by mid-century the age of the universal man was passing away. His work on the calculus of finite differences, which was the focus of his early papers published in the *Philosophical Transactions*, was no doubt his principal contribution to mathematics It was also explored in his appendix in the translation of Lacroix, a work which was duly noted by P. G. Tait (1831–1901) 'as one of the most charming mathematical works ever written, everywhere showing power and originality, as well as elegance. In all these respects it far surpasses his subsequent mathematical writings, excellent as are many of them'.[49] Herschel's work on finite differences places him among the first generation in the development of algebra in Britain. Duncan Gregory and other writers on the calculus of operations in the *Cambridge Mathematical Journal* were the next generation.[50] Herschel's difference operator Δ_h was one algebraic operator in the calculus of operations defined by the 'separation of symbols' adopted by many second wave of British mathematicians in the 1840s. Setting off on his own mathematical career, mathematicians like Arthur Cayley were brought up in an environment where the calculus of operations was an active field of research. In the 1840s, Cayley set in train his great invariant theory programme, borrowing his technique from Herschel's in dealing with the calculus of differences.[51]

Despite this influence, Herschel could not keep up with rapid changes in science made in the 1850s and 1860s. He was from an earlier age, where people like himself and Whewell flourished in their 'omnipotence'. By mid-century, specialists in mathematics in Britain were emerging, people like Cayley and Sylvester, who for all their lives restricted themselves to delving into a single subject.

Notes

1 Slough was subsequently transformed from an English village to a town of 'brutalist concrete' prompting the famous lines from Poet Laureate John Betjeman (1906–84): 'Come, friendly bombs, and fall on Slough.'

2 M. Fisch, *Creatively Undecided: Towards a History and Philosophy of Scientific Agency* (Chicago: University of Chicago Press, 2017), 143.

3 Alexander Rogers to John Herschel, 18 October 1833, Royal Society of London HS (henceforth RS:HS) 14.417; 'Herschel2172', in *Epsilon: The Sir John Herschel Collection*, accessed on 10 February 2023, https://epsilon.ac .uk/view/herschel/letters/Herschel2172.

4 H. Becher, 'Woodhouse, Babbage, Peacock and Modern Algebra', *Historia Mathematica*, 7 (1980): 389–400.

5 John Toplis, *A Treatise upon Analytical Mechanics* (Nottingham: H. Barnett, 1814).

6 S. E. De Morgan, *Memoir of Augustus De Morgan* (London: Longmans Green, 1882), 41.

7 G. Peacock, 'Report on Certain Branches of Analysis', *Report on the Third Meeting of the British Association for the Advancement of Science* (1833): 185–352, on 285 and 296.

8 John Herschel to William Herschel, 1 December 1812, Harry Ransom Center, University of Texas at Austin (hereafter HRC), quoted in P. Enros, 'The Analytical Society (1812–1813) Precursor of the Renewal of Cambridge Mathematics', *Historia Mathematica*, 10 (1983): 24–47, on 30.

9 John Herschel, 'Analytical Formulae for the Tangent, Cotangent, &c.', *Nicholson's Philosophical Journal*, 31 (February 1812): 133–36; John Herschel, 'Trigonometrical Formulae for Sines and Cosines', *Nicholson's Philosophical Journal*, 32 (March 1812): 13–16.

10 This was the place and date of the founding of the Analytical Society settled on by Enros in Enros, 'The Analytical Society (1812–1813)', 27. There is some disagreement on the date of the inaugural meeting. See, for instance, later scholarship in M. Fisch, 'The Making of Peacock's *Treatise on Algebra*: A Case of Creative Indecision', *Archive for the History of Exact Sciences*, 54 (1999): 137–79, on 137 n. 2. Further work on the Analytical Society includes H. Becher, 'Radicals, Whigs and Conservatives: the Middle and Lower Classes in the Analytical Revolution at Cambridge in the Age of Aristocracy', *British Journal for the History of Science*, 28.4 (1995): 405–26; and M. V. Wilkes, 'Herschel, Peacock, Babbage and the Development of the Cambridge Curriculum', *Notes and Records of the Royal Society of London*, 44 (1990): 205–19. A retrospective view of the Society's establishment is given in C. Babbage, *Passages from the Life of a Philosopher* (London: Longman, Roberts & Green, 1864).

11 F. Maule to C. Babbage, 16 January 1813, British Library, BLK Add. 37, 182, f 3. Quoted in Anthony Hyman, *Charles Babbage: Pioneer of the Computer* (Oxford: Oxford University Press, 1982), 25.

12 Quoted in Wilkes, 'Herschel, Peacock, Babbage', 214, and Enros, 'The Analytical Society', 30–31.

13 Cotes's theorem is concerned with expressing $x^n \pm a^n$ into linear and quadratic factors and can be cast into geometric form involving conics.

14 Brinkley was an influence on Herschel's early mathematical work, both with respect to Cotes's theorem and the calculus of differences. He was senior wrangler at Cambridge in 1788 and pursued a career in Ireland as an astronomer. He was known to Herschel as a member of the Astronomical Society and served as its president 1831–33. His two papers related to Herschel's work were John Brinkley, 'A General Demonstration of the Property of the Circle discovered by Mr Cotes deduced from the Circle Only', *Transactions of the Royal Irish Academy*, 7 (1800): 151–59; and 'An Investigation of the General Term of an Important Series in the Inverse Method of Finite Differences', *Philosophical Transactions of the Royal Society*, 97 (1807): 114–32.

15 John Herschel, 'On a Remarkable Application of Cotes's Theorem', *Philosophical Transactions of the Royal Society of London*, 103 (1813): 8–26.

16 W. Whewell to his father, 17 February 1813, in Mrs. Douglas Stair, *The Life and Selections from the Correspondence of William Whewell, DD. Late Master of Trinity College Cambridge* (London: Kegan Paul, 1881), 10.

17 John Herschel to Edward Thomas Ffrench Bromhead, 19 November 1813, quoted in Enros, 'The Analytical Society', 41.

18 Enros, 'The Analytical Society', 33–40.

19 John Herschel to C. Babbage, 12 January 1812, quoted in Enros, 'The Analytical Society', 34.

20 Quoted in Enros, 'The Analytical Society', 40.

21 Quoted in Hyman, *Charles Babbage*, 27.

22 For the philosophical underpinings supplied by the Cambridge quartet of Herschel, Babbage, Peacock, and Whewell, see Fisch, *Creatively Undecided*.

23 W. Whewell, 'Transactions of the Cambridge Philosophical Society. Science of the English Universities', *British Critic*, 9 (1831): 71–90, on 85. This is quoted in Enros, 'The Analytical Society', 37, where he identifies the anonymous reviewer as Whewell.

24 Herschel was entered at Lincoln's Inn as a twenty-one-year-old on 21 January 1814 (Folio 45, *Admissions Register vol. 2 1800–1893* [1896]) but left the Law without being called to the Bar.

25 John Herschel, 'Considerations of Various Points of Analysis', *Philosophical Transactions of the Royal Society of London*, 104 (1814): 440–68.

26 See Herschel, 'Consideration of various Points of Analysis', 441. The separation of symbols was a distinctive feature of the analytical method, not least because the resulting notation enabled theorems to be expressed concisely. In present day mathematics, Δ_h is called the forward difference operator and is extensively used in numerical analysis. For the importance

of the calculus of operations in British mathematics, see E. Koppelman, 'The
Calculus of Operations and the Rise of Abstract Algebra', *Archive for
History of Exact Sciences*, 8 (1971): 155–242.

27 John Herschel to C. Babbage, 18 December 1815, RS:HS 2.51,
'Herschel5102', in *Epsilon: The Sir John Herschel Collection*, accessed on
10 February 2023, https://epsilon.ac.uk/view/herschel/letters/Herschel5102.
Quoted in C. A. Ronan, 'John Herschel (1792–1871)', *Endeavour*, 16 (1992):
178–81, on 179.

28 John Herschel, 'On the Development of Exponential Functions, Together
with Several New Theorems Relating to Finite Differences', *Philosophical
Transactions of the Royal Society of London*, 106 (1816): 25–45.

29 W. R. Hamilton treated Herschel's Theorem as a basis for further research
in 'On Differences and Differentials of Functions of Zero', *Transactions of
the Royal Irish Academy*, 17 (1837): 235–36. It is also treated as a topic in A.
De Morgan's compendious *Differential and Integral Calculus* (London:
Baldwin and Cradock, 1842), 307–8. By the time G. Boole's *Treatise on the
Calculus of Finite Differences* (London: Macmillan, 1860) was published,
Herschel's Theorem was utilised but assumed known to its readers. It has
recently been rediscovered by L. Fekih-Ahmed. See L. Fekih-Ahmed, 'On
Two Applications of Herschel's Theorem', 5 May 2012, accessed on
10 February 2023, https://arxiv.org/pdf/1205.1104.pdf.

30 John Herschel to C. Babbage, 10 September 1816. Quoted in A. M. Clerke,
'John Frederick William Herschel (1792–1871)', *Dictionary of National
Biography*, 26 (1891), on 263.

31 John Herschel to C. Babbage, 10 October 1816, RS:HS 2.68, 'Herschel5119',
in *Epsilon: The Sir John Herschel Collection*, accessed on 10 February
2023, https://epsilon.ac.uk/view/herschel/letters/Herschel5119. Quoted in
Ronan 'John Herschel', 179.

32 John Herschel, 'On Circulating Functions, and on the Integration of a Class
of Equations of Finite Differences into which They Enter as Coefficients',
Philosophical Transactions of the Royal Society of London, 108
(1818): 144–68.

33 Enros, 'The Analytical Society', 39.

34 Lacroix was a dedicated writer of mathematical textbooks. His *Traité
élementaire* was his most successful and was reprinted throughout the
nineteenth century. (By 1881, a ninth edition had been published.)
Translations were made into Portuguese (1812), German (1817), Polish
(1824), and Italian (1829). The English translation appeared in 1816. See J.
M. CdM. Domingues, *The Calculus according to S. F. Lacroix* (PhD thesis,
Middlesex University, 2007) for an extensive study of this text and
its historiography.

35 The appendix spanned pages 465–569. Herschel's translational work involved the section on the integral calculus (with Peacock) and notes N, O, P, and Q, 682–711.

36 D M. Peacock, *A Comparative View of the Principles of the Fluxional and Differential Calculus* (Cambridge, 1819), 85. See also Becher, 'Radicals, Whigs and Conservatives', 416–20, and Enros, 'The Analytical Society', 37–38.

37 J. Challis, *Remarks on Cambridge Mathematical Studies and their Relation to Modern Physical Science* (Cambridge: Deighton, Bell, 1875), 5.

38 John Herschel, *A Collection of Examples of the Applications of the Calculus of Finite Differences*; G. Peacock, *A Collection of Examples of the Applications of Differential and Integral Calculus*; and C. Babbage, *Examples of the Solutions of Functional Equations* (Cambridge: J. Deighton, 1820). G. Peacock to John Herschel, 18 February 1822, quoted in Jonathan Topham, 'A Textbook Revolution', in Marina Frasca-Spada and Nick Jardine, eds., *Books and the Sciences in History* (Cambridge: Cambridge University Press, 2000): 317–37, on 331.

39 Herschel was awarded a second Copley Medal (1847) for the astronomical observations carried out during his stay at the Cape of Good Hope 1834–38.

40 W. Whewell to John Herschel, 23 November 1827, in Isaac Todhunter, *William Whewell, D.D.: An Account of His Writings with Selections from His Literary and Scientific Correspondence*, vol. 2 (London: MacMillan, 1876), 86.

41 John Herschel, 'Mathematics', *Edinburgh Encyclopaedia*, 13 (1830): 358–83, on 359.

42 'Presidential Address', *Report of the Fifteenth Meeting of the British Association for the Advancement of Science* (1845).

43 The relevant section in John Herschel, 'Quételet on Probabilities', *Edinburgh Review*, 92 (July 1850): 20–21. Reprinted in John Herschel, *Essays from the Edinburgh and Quarterly Review* (1857): 399–400. For the suggestion that Maxwell read and was influenced by this article, see S. Brush, *The Kind of Motion We Call Heat*, vol. 2 (North Holland: Elsevier Science, 1976), 342.

44 John Herschel, 'On the Algebraic Expression of the Number of Partitions of Which a Given Number is Susceptible', *Philosophical Transactions of the Royal Society of London*, 140 (1850): 399–422. In this work, Herschel anticipated Arthur Cayley, 'Researches on the Partition of Numbers', *Philosophical Transactions of the Royal Society of London*, 146 (1856): 127–40. Herschel used results from his early papers: 'On the Development of Exponential Functions', *Philosophical Transactions of the Royal Society of London* 106 (1816): 25–45; and 'On Circulating Functions', *Philosophical Transactions of the Royal Society of London*, 108 (1818): 144–68.

45 John Herschel, 'On the Formulae Investigated by Dr. Brinkley for the General Term in the Development of Lagrange's Expression for the Summation of Series and for Successive Integrations', *Philosophical Transactions of the Royal Society of London*, 150 (1860): 319–21.

46 Herschel, 'On Circulating Functions'. See G. E. Andrews, 'Partitions', in R. J. Wilson and J. J. Watkins, eds., *Combinatorics: Ancient and Modern* (Oxford: Oxford University Press, 2013): 205–29.

47 J. J. Sylvester to John Herschel, 5 March 1867. RS:HS 17.164; Herschel is not on record as joining the London Mathematical Society according to a definitive list of former LMS members issued c.1900. (John Heard, private communication, 8 January 2023).

48 John Herschel, 'On a New Projection of the Sphere', *Proceedings of the Royal Geographical Society*, 3 (1859): 174–77. The mathematics is presented in John Herschel, 'On a New Projection of the Sphere', *Journal of the Royal Geographical Society*, 30 (1860): 100–106.

49 P. G. Tait, *Proceedings of the Royal Society of Edinburgh*, 7 (1871–72): 544–45.

50 On the relationship between Peacock and his 'science of symbols' and that of symbolic algebraist D. F. Gregory exploring the calculus of operations and Cayley and the development of abstract algebra, see L. M. Verburgt, 'Duncan F. Gregory, William Walton and the Development of British Algebra: "Algebraical Geometry", "Geometrical Algebra", Abstraction', *Annals of Science*, 73 (2016): 40–67.

51 A. Cayley, 'On Linear Transformations', *Cambridge and Dublin Mathematical Journal*, 1 (1846): 104–22, on 104.

STEPHEN CASE

3

John Herschel's Astronomy

Despite John Herschel's extensive work in the fields of chemistry, optics, geology, mineralogy, and the philosophy of science, it was primarily as an astronomer that he was recognized during his lifetime and remembered after his death. Herschel's astronomical endeavors can be summarized as *establishing* and *extending* the astronomical projects of his father William Herschel (1738–1822) and his aunt Caroline Herschel (1750–1848). By *establishing*, Herschel brought an observational and mathematical rigor to his father's observations that transformed them from the results of an individual with unique instrumentation to data useful and accessible to the wider astronomical community. By *extending*, Herschel continued the observational program of his father and aunt, revisiting his father's observations, updating his catalogs of double stars, nebulae, and star clusters, and extending the Herschelian project to the skies of the southern hemisphere. Yet the observational aims and methods of William and Caroline Herschel were markedly different from the astronomy being pursued by most other astronomers during their lifetimes. To understand John Herschel's long astronomical career and influence, an overview of the scope and aims of the dominant form of astronomy during this period is needed that can provide the background to the work of his father and aunt and the context of his own.

The Context of Herschelian Astronomy

Prior to the work of the Herschels, observational astronomy was largely *positional astronomy*. Astronomers, both amateur and professional (employed by government or university observatories, or even in the pay of rich amateurs), used telescopes mounted along the meridian in permanent observatories to time stellar culminations. From the data of altitude and time of culmination and after a mathematically rigorous process of correcting for factors such as atmospheric refraction, aberration, and precession, the position in right ascension and declination of observed stars could be determined to a high degree of accuracy. The products that emerged from this observational and calculational process were catalogs of star positions. Such "ledgers full of stars" were rather more in the tradition of precise accounting books than resources for planetary or sidereal discovery. They were not seen as a means of producing new physical knowledge but rather providing more exact measurements of the stellar background against which other important motions, such as the motion of the moon or planets, could be precisely determined.[1]

Instrumentally, positional astronomy utilized refracting telescopes in conjunction with clocks (to measure time of culmination of observed star) and meridian circles (to measure telescope angle and thus altitude of observed star). This form of astronomy was essential for navigation. Latitude on the Earth's surface was measured from the observed altitude of specific stars, but longitude could not be measured directly and posed a significant technical challenge. One method was measuring "lunars," which utilized the distance of the moon from certain stars to determine longitude. For this method to be effective, accurate catalogs of star positions were needed. For a maritime, mercantile nation like Great Britain, positional astronomy was closely linked with trade, commerce, and naval power. The precision of star catalogs that emerged from national observatories signified the country's technical and instrumental superiority, while catalogs from British colonies such as the Cape of Good Hope or New South Wales served the projects of empire. By the nineteenth century, astronomy was so much equated with this approach that the *Penny Cyclopedia* of 1843 could enjoin readers to "see *transit instrument*" in place of its entry for "observatory."[2]

Positional astronomy also served physical astronomy: applying Newtonian gravity to the solar system required the position of planets, the moon, and comets measured in relation to accurate star positions. In this sense, positional astronomy provided a background against which the physical astronomy of the solar system could be worked out. The stars, as John Herschel put it, were the "landmarks of the universe," though astronomers were not generally interested in sidereal objects themselves.[3] The catalog of Charles Messier (1730–1817) serves in this respect as the exception that proves the rule: Messier composed his list of nebulous objects (including objects today characterized as galaxies, nebulae, or star clusters) so he and other observers could more easily avoid these bits of observational noise in their pursuit of comets.

In this context, the revolutionary nature of the work of William and Caroline Herschel becomes apparent. William, fascinated by the objects he observed through his large instruments, initiated a program of astronomy focused on the sidereal universe for its own sake. When William, a Hanoverian immigrant and musician, started casting mirrors and creating reflecting telescopes, he began an observing campaign beyond the remit of positional astronomy both in aims and instrumentation. By 1783, inspired by Caroline's observations made while searching for comets, he had commenced observational "sweeps" for nebulae and star clusters with a large twenty-foot reflecting telescope of his own construction and design. The size of his telescopes, which made them the most powerful in use at that time, brought hundreds of new objects to view. Between 1786 and 1802, William published three catalogs of nebulae and star clusters, containing 2,500 objects, most of which had never been observed before.[4] William used his observations to create a natural history of the heavens, speculating on the nature of nebulae and their condensation into stars. He also conducted surveys of double stars and worked to organize stars into lists by brightness as a means of searching for variable stars. By returning to observe double stars later, he discovered that many exhibited rotational motion, implying gravitational attraction between their components and supporting the extension of Newtonian gravity throughout the universe. In an 1803 paper, Herschel proposed that some of these stars were true *binary* pairs.[5]

Though William's observing program had slowed by the time John Herschel was born, William's only son grew up in the literal shadow of the telescope, in particular the massive forty-foot reflector that William built with financial support from the king and that towered over their home in Slough, outside of London. Some of John's earliest memories were of finding his aunt Caroline in her workroom, where she would entertain him by showing him illustrations of constellations in the star catalog of John Flamsteed (1646–1719).[6] John's first recorded observation was of the shape of Saturn as seen through his father's telescope in 1808, made when he was fifteen.[7] Apart from witnessing his father's observations and the steady stream of domestic and international visitors who made the trip from London to see his instruments and discuss astronomy, John also witnessed Caroline's patient dedication in organizing and reducing the data generated by William's observations, translating it through her work into publishable form. This painstaking attention to detail would serve Herschel well in his own work, and his astronomical career combined both the observational practice of his father and the data processing skills of his aunt.

A Reluctant Astronomer

Though the work of William and Caroline Herschel represented a shift in focus to objects beyond the solar system, there was no one else in the astronomical community to follow up on their observations or help establish their revolution. No one besides William had the instruments that allowed such observations, and he did not pass on his techniques for creating the mirrors essential to his telescopes or take on apprentices – except for one. John Herschel inherited the unique astronomical project of surveying the sidereal heavens and the telescopes and training to carry it through. Being an astronomer, however, was something the younger Herschel originally resisted. John had been afforded the best primary education available through the substantial fortune inherited by his mother, Mary Pitt Herschel (1750–1832), including tutoring in advanced mathematics. This training was so effective that, when Herschel enrolled at Cambridge, he was appalled by the poor state of mathematical instruction at the university. He began his scholarly career not in astronomy but intending to reform the practice and theory of mathematics in Britain.[8] After graduating, Herschel briefly studied law before pursuing chemistry and optics and only finally agreed to become his father's apprentice after a stint tutoring at Cambridge and a nervous breakdown. Though Herschel would go on to publish numerous nebulae and double star catalogs, his work in astronomy always went hand-in-hand with his involvement in scientific societies, experimental work in chemistry and optics, and the writing of popular books and encyclopedia articles.

Herschel's career in astronomy began in the fall of 1816, when he started working with his father to restore the twenty-foot reflector and began keeping an observing journal, but actual observing proceeded in a desultory manner as he traveled extensively in Europe and became involved in the Royal Society of London and the forming of the Astronomical Society of London. This latter society was established in 1820 to support and standardize observations of a group of middle-class amateurs and businessmen who wanted to bring the structure and rigor of finance to astronomical practice and felt the Royal Society was too conservative for their reforms.[9] The group included Herschel's close friend and fellow mathematician Charles Babbage (1791–1871) as well as London professionals such as the accountant Francis Baily (1774–1844) and physician James South (1785–1867). In his role as foreign secretary to the new society, Herschel leveraged the credibility of his family name to create an international correspondence network for gathering, publishing, and disseminating astronomical data. Though the Astronomical Society quickly earned the ire of Joseph Banks (1743–1820), the autocratic president of the Royal Society, the new society brought new rigor and standardization to astronomy as well as a way for middle-class individuals from professional (nonscientific) backgrounds to contribute to the field.

Whereas his father had remained on the margins of scientific society, Herschel worked to make London a center of astronomical observation and data as much as it was a center of trade and mercantilism. Similarly, whereas William was satisfied working independently, Herschel worked to create a unified body of observers with standardized aims and procedures, bringing a cohesive direction to the disconnected work of the previous generation of amateur observers. Herschel also cultivated relationships between astronomers and the British government and admiralty. As a member of the Board of Longitude, he worked to integrate instrumentation and instructions for observations into naval missions, reform the *Nautical Almanac* to make it more useful to astronomers, advise on the establishments of observatories in the southern hemisphere, and apply his knowledge of optical theory and telescope design to testing the quality of glass produced by London craftsmen. In the midst of this work building institutional structures to support astronomy, Herschel's own observational practice started slowly and in London, rather than at his father's side at Slough, where he could remain connected to the social and scientific culture of the metropolis.

Double Star Astronomy

John Herschel's first astronomical research program, commencing in 1821, was observing double stars with James South from the roof of South's home in London. South had devoted himself to astronomy after gaining financial independence through marriage and retiring from his physician's practice. His notoriously stormy disposition resulted in feuds with colleagues and instrument makers that ultimately alienated many of his friends, including Herschel, but initially the collaboration between the two was fruitful. South had two high-quality refracting telescopes with equatorial mounts. These instruments, in contrast to meridian-mounted instruments in permanent observatories or Herschel's large reflectors, could track an object across the sky. Though they lacked the power of William's telescopes, they were thus ideal for finding particular celestial objects and holding them in view so delicate measurements of the objects themselves (not simply their positions or descriptions) could be made. Herschel and South put these instruments to use creating catalogs of double stars and revisiting William Herschel's earlier discoveries.

Double stars were an example of the intersection of observation, instrumentation, and mathematical analysis leading to physical insights regarding the sidereal universe. It had been realized since the early days of telescopic observations that some stars appearing as single stars to the naked eye actually consisted of two or more stars in close proximity. Galileo had proposed using such stars as a means of detecting parallax, the observational proof of the Earth's motion about the Sun. If such double stars were simply line-of-sight doubles, distant stars that appeared close together in the sky due only to our perspective, then as the Earth orbited the Sun the relative position between the two components of the double star should shift annually. It was with the thought of detecting parallax by this means that William Herschel began his own double star project, using his large telescopes to add to the number of known double stars. He published his catalogs of double stars, containing over seven hundred new objects, in 1782 and 1785.[10]

Ironically, William's observations left the field less accessible to subsequent observers. Revisiting some of the double stars from his original catalog years later led to his discovery that their component stars had indeed shifted but not in a way that could be explained by parallax. Instead, it appeared that some of William's double stars were true *binary* doubles, stars bound gravitationally to one another and orbiting around a common center of gravity. This possibility had been originally proposed by John Michell (1724–63) in a *Philosophical Transactions* article prior to William's discovery in which Michell used statistical analysis to show the number of known double stars was unlikely if they were all chance line-of-sight doubles not due to true physical associations.[11] William's observations, however, were the first empirical evidence for this association and thus for the extension of Newtonian gravity to the sidereal universe. John Herschel considered this the greatest of his father's discoveries – if not "one of the greatest ever made by man" – beside which William's 1781 discovery of Uranus was "but a trifle."[12]

Yet this discovery problematized subsequent double star observations. With the introduction of true binaries into the celestial taxonomy, there was no way for future observers to know whether any particular double star observed was an optical, line-of-sight double, and thus useful for the ongoing search for parallax, or a true binary. One way to address this was comparing current positions and distances of the component stars in a particular double star pair with previously recorded observations. For the vast majority of double stars, however, the only observations were those of William Herschel's catalogs. Unfortunately, William's double star measurements were not always sufficiently precise, and his catalogs, like his nebulae catalogs, were poorly ordered for subsequent observers. To make double stars useful targets for the growing community of astronomers, William's double stars needed to be systematically re-observed and re-measured. This would simultaneously provide a more exact baseline for future double star observations and, by comparisons with previous measurements, indicate which pairs were likely line-of-sight and which likely binary doubles. This was the project Herschel and South embarked on together.

The relevant data for determining double star motion was careful measurement of the distance between the two component stars and the angle of a line joining them measured from some reference direction (Herschel used north). Besides South's equatorial telescopes, Herschel and South had additional precision tools to make this possible: micrometers. A micrometer consisted of two thin parallel wires placed within the focal plane of the telescope. When illuminated with a small lantern, the wires were visible in the telescope's field of view. (This need for illumination meant micrometers were most useful in observing bright objects such as stars and planets.) The two wires could be rotated and the distance between them adjusted. By varying these two parameters, the distance between the components of a double star and the angle of their separation could be measured to a high degree of precision. In 1824, Herschel and South published a catalog of 380 double stars based on their observations from 1821 to 1823.[13]

Herschel's double star astronomy would soon become an adjunct to his project of extending and completing his father's sweeps for nebulae, but it showed important themes in Herschel's approach that would continue throughout his astronomical career. Measuring double stars was at the frontier of what high quality telescopes could hope to accomplish. Large telescopes were required to resolve very tight doubles, and viewing conditions and instrument quality meant measurements could differ significantly from observer to observer or from night to night. Making such observations *reliable* became central to Herschel's scientific methodology. For double star observations, this meant building a network of observers and developing a means of reporting measurements consistently and weighing averages based on confidence of data. Throughout the 1820s, Herschel encouraged other observers to begin measuring double stars by his method and printed thousands of skeleton forms to record data in a standard form.

Double stars also provided a means of integrating mathematical analysis into astronomy. Herschel first developed a method of determining parallax (though never put it into practice) based on change in the angular position of double star components, which he knew from observing experience was more reliably determined than distance.[14] As more close double stars were discovered, he developed a method of calculating binary star orbits. This differed from the methods already utilized by Félix Savary (1797–1841) and Johann Franz Encke (1791–1865) in that, whereas their methods were highly analytical, Herschel's experience as an observer led him to believe double stars could not be treated with the rigor of planetary positions. Herschel's method, like his method for parallax, was dependent on the more robust angular measures and utilized graphical techniques involving fitting curves by hand.[15]

Though Herschel began his sweeps for nebula by 1824, discovering and measuring double stars remained a common thread alongside this work. He published a second catalog of doubles in 1826, containing over three hundred new doubles, followed by a third and fourth in 1829, bringing the total to nearly one thousand objects discovered in the course of his sweeps.[16] Herschel purchased one of South's refracting equatorials in order to measure double stars more precisely and published a catalog of micrometrical measures of doubles in 1833.[17] By the time he left for the Cape of Good Hope, he had established a network of observers both abroad, including Giovanni Battista Amici (1786–1863) and Wilhelm Struve (1793–1864), and in England, with observers such as William Rutter Dawes (1799–1868), William Henry Smyth (1788–1865), and John Russell Hind (1823–94). He continued his double star work during his years observing at the Cape of Good Hope, including a catalog of southern doubles as part of his monumental *Cape Results*, published upon his return to England. Near the end of his life, Herschel returned to double stars, compiling all known observations into a general catalog containing over ten thousand double stars.[18] Today, Herschel's double star discoveries are still signified in many modern star atlases by a lower-case *h* proceeding the star's number from Herschel's catalogs.

Double stars were an important and enduring aspect of Herschel's career, establishing William's pioneering work, both by increasing the utility of William's original observations for other observers and by using his own influence to encourage observations of double stars among other astronomers. Double stars became a primary way in which non-positional, extra-meridional astronomy yielded physical data regarding the sidereal universe. The first double star orbits were based on Herschel's observations and provided a direct means of measuring stellar masses. Though Herschel never claimed to have *proven* the gravitationally bound nature of binary stars, consistently contributing that discovery to his father, William's data was not enough to verify this claim. John Herschel's subsequent work made this possible.

Nebulae

By 1822, Herschel had begun his double star work with South but not yet committed to taking up his father's sweeps for nebulae. It took tragedy to bring him to begin that immense undertaking. After a trip to France in 1819 and a long expedition through the Alps with Babbage in 1821, Herschel left England the following year for a brief tour of the Netherlands. He was hesitant to depart because of his aging father's health, and unfortunately his fears proved well-founded. Herschel returned to London to news that his father had died almost a week prior and that he had missed the funeral. It was certainly with a sense of filial piety and as a means of honoring his father's legacy that Herschel resumed William's nebulae observations within a month of burying his father in their parish church in Slough.

William Herschel's most enduring astronomical project had been his discovery of thousands of new sidereal objects, published in his three large catalogs. In these catalogs, objects were classified by appearance, and positions were given with respect to reference stars with distance estimated using diameter of the telescope's field of view. William's observations led him to speculate on the nature of nebulae, first believing there was a distinction between nebular mist and stars; then, as he was able to resolve more nebulae into discrete stars, that all nebulae were only unresolved stars; and finally, reverting back to his first view that there were indeed regions of irresolvable nebulosity in the heavens. These ideas led him to a theory of the evolution of sidereal systems, with William ultimately concluding that nebulosity (a "luminous fluid") collapsed to form stars. He published this final theory in 1814, when John Herschel was a student at Cambridge.[19]

William Herschel's "natural history" of the heavens raised the question of whether it was possible to observe changes in nebulae over time. If nebulae formed stellar systems, then it could be possible to see stars and even planetary systems in the process of formation. Because the appearance of nebulae varied significantly due to viewing conditions, however, it was notoriously difficult to determine whether changes or condensation in nebulosity had indeed occurred. As with double star positions, appearance of nebulae could change from night to night. Standardized observations were needed. Unfortunately, William's catalogs, in which objects were organized in order of discovery rather than position in the sky, were poorly structured for use as observing guides. John Herschel's observations of nebulae, like his double star project, was an attempt to both make William's original observations accessible to other observers and revisit these objects to compare their current appearance with William's descriptions.

In this project, Herschel was greatly aided by the continuing work of his aunt Caroline. After William's death, Caroline had returned to her original home of Hanover in Germany, where she reorganized her brother's catalogs and created an updated list of all his objects in order of right ascension. When Herschel urged her to publish her work, she refused, maintaining she had created it for his personal use as a tool to revisit and complete William's sweeps. Herschel acknowledged he would not have begun the project of revisiting his father's nebulae and clusters without Caroline's catalog and that it should be taken as the "groundwork" of the entire project.[20] Herschel saw his resumption of his father's sweeps as establishing an empirical baseline for nebulae and star clusters, providing accurate positions for each object so observers could find them and a sufficient description so subsequent observers would be able to determine whether meaningful physical changes had taken place. The catalog that resulted, covering observations Herschel made from 1825 to 1833, was published in the Royal Society's *Philosophical Transactions* and described by the Astronomer Royal George Airy (1801–92) as the beginning of this science, the "first accurate account" of the sidereal universe, placing observations of nebulae on an empirical footing.[21]

The Slough catalog, besides including nearly five hundred new nebulae and star clusters, formed a nearly complete record of sidereal objects known through 1833. Its strength was in the absolute positions provided for each object (to a tenth of a second in right ascension and one minute in declination, which Herschel measured as north polar distance) and the detail of object descriptions. William had focused on discovery; John Herschel focused on precision. His observations of nebulae led him to distinguish between irregular (irresolvable) nebulae that likely did congeal into stars and regular (in principle resolvable) nebulae that were groups of stars, though he remained circumspect in his theorizing and insisted the matter could not be resolved until changes within a nebula had been observed. This could be solved with more observers watching nebulae, something he hoped the catalog would allow, or by more nebulae to observe, something that would require new skies. Fortunately for Herschel, the southern hemisphere offered vast tracts of the heavens untouched by any instrument of the caliber of his reflector.

Herschel had been considering extending his surveys to the southern skies for some time, and with the death of his mother in 1832 he made the decision to travel to the Cape of Good Hope in South Africa. Though the Cape had a permanent government (meridional) observatory and Herschel was offered passage on a navy ship, he insisted on the entire operation remaining a private (and privately funded) endeavor. In 1834, he took with him his young wife, Margaret Brodie (née Stewart, 1810–84), and their three young children, as well as a mechanic named John Stone to help with the operation of the telescope, and a nurse, who died soon after their arrival. At the Cape, like at Slough, Herschel worked from home, a house he called Feldhausen, with the twenty-foot reflector and the seven-foot refractor set up on the lawn.

By the time Herschel left England for the Cape, he was already well-known for his astronomical work as well as his mathematics and his popular writings. He had published encyclopedia articles on sound and light, both substantial treatises. The work on light in particular was considered the most significant treatment since Newton's *Optics* and provided a survey of both wave and corpuscular theories.[22] His 1830 work on scientific methodology, *A Preliminary Discourse on the Study of Natural Philosophy*, made him the spokesperson for the scientific vocation in Britain, while his 1833 textbook, *Treatise on Astronomy*, brought his astronomical discoveries to an even wider audience. It was thus as not only an astronomer but a well-known scientific personality that Herschel left for his Cape expedition.

Herschel's four years at the Cape resulted in important revisions of many of his astronomical concepts. In particular, the Large and Small Magellanic Clouds offered a compelling display of stars, star clusters, and nebulosity in apparent juxtaposition, strengthening Herschel's belief in true nebulosity. Another object of interest was the star known as η Argus (today η Carina) and the nebula surrounding it. The stars interspersed within this nebula provided Herschel a chance to map nebulosity with meticulous precision and to establish a baseline for measuring any possible future movement or change. Herschel painstakingly charted hundreds of stars clustered within the nebula, recording the state of the nebula with as much precision as pre-photographic methods allowed (see Chapter 6). The drawings Herschel made at the Cape became the data by which Herschel would test claims to changes in the nebula decades later.[23]

During his time at the Cape, Herschel performed nearly four hundred sweeps, continuing the numbering scheme he had used at Slough. This survey made him the first astronomer to closely examine the entire heavens by telescope. Though the Slough catalog had been published, like William's nebulae catalogs, in the *Philosophical Transactions*, Herschel published his southern catalog as part of a larger work, an immense volume sponsored by Hugh Percy (1785–1847), third Duke of Northumberland, containing the results of all his observations in the southern hemisphere and usually referred to as the *Cape Results*. This work was not published until 1847, almost a decade after his return from South Africa, and contained, in addition to the nebula catalog, chapters on observations of the Sun, the magnitudes of stars, Halley's comet, and double stars.[24]

The catalog of nebulae and star clusters in the *Cape Results*, which included over 1,200 new objects, represented the completion of the Herschelian project of sidereal astronomy extended to the entire heavens. In the catalog, Herschel offered a classification scheme building on his father's earlier work, in which objects were divided into regular and irregular nebulae and irregular clusters. Herschel classified most of the objects he discovered as regular nebulae, which he then evaluated on five criteria (magnitude, brightness, roundness, condensation, and resolvability), ranking each criterion by five grades. Ultimately, Herschel hoped to digitize the scheme by transforming each object's description into a sequence of numbers. This method of standardizing the properties of nebulae never caught on, possibly because by the time the *Cape Results* were published larger telescopes such as that of William Parsons (1800–67), the third Earl of Rosse, were providing new, more detailed views of objects that had up until then been visible only through Herschel's large reflectors.

Regarding the nature of the nebulae, Herschel's time at the Cape gave him further grounds for doubting that some nebulae were in principle resolvable and extended his ideas regarding the nature of nebulosity itself. If luminous matter could be composed of discrete objects, analogous to droplets of water in a cloud or fog, there might be no real distinction between clusters of stars and nebulae as clouds of minute luminous matter. In fact, this idea would gain additional strength through observations not of nebulae but of a single remarkable star.

Variable Stars and Stellar Magnitudes

While the nebulosity around η Argus provided a groundwork for mapping nebulae, it was η Argus itself that caused Herschel the most surprise at the Cape. On the evening of December 16, 1837, he noticed that the star, which was normally of fourth magnitude, had increased so markedly in brightness that he was for a time at a loss to identify it. It continued to increase, eventually rivaling Rigel to appear as the brightest star in the sky.[25] Its unexpected brightening and subsequent dimming turned Herschel's attention to the problem of variable stars. Prior to this, variable stars were assumed to come in two types: periodic variables, such as Algol, that were explained by dark spots on their surfaces regularly turning toward Earth as the star rotated, and non-periodic variables, or novae, that appeared and then dimmed slowly and were clearly eruptions of unknown cause. η Argus offered a new type of variable midway between the two: a star that erupted and dimmed without periodicity and without disappearing like a nova.

Herschel was already engaged in creating lists of stars at the Cape organized by relative magnitude, but the eruption of η Argus lent a new urgency to this study and caused Herschel to speculate on the possibility of "super-atmospheric clouds" that could thicken or clear, obscuring and revealing stars, as a possible mechanism for η Argus' variability.[26]

Herschel's experience with the sudden and unexpected variability of η Argus gave him a renewed interest in quantifying stellar magnitudes. He had already been following the process originated by his father of organizing lists of stars in sections of the sky in order of their apparent brightness. This allowed stellar variability to be detected by change in the ordering of these lists. Without instrumentation to measure star magnitudes directly, this comparative method was a means of bringing stellar variability within the reach of amateur and naked eye observers. η Argus convinced Herschel there was much to be discovered regarding the physical mechanisms of changes in magnitude and that variable star observations represented an important and neglected field of astronomy. While at the Cape, he developed a photometer that allowed quantitative measures of stellar brightness, but in practice Herschel preferred the method of comparative star lists.[27]

After his return from the Cape, naked-eye comparisons of stellar magnitudes were Herschel's only continuing astronomical observations. He urged observers to pay closer attention to stellar magnitude, believing the periodic variability of even such bright stars as α Cassiopeiae and Betelgeuse had been overlooked.[28] When, in 1849, Herschel edited a manual issued to navy officers for making scientific observations during sea voyages, Airy wrote the chapter on astronomical observations but Herschel used his editorial oversight to include lists of magnitudes for northern and southern hemisphere stars and instructions on creating magnitude lists that would stitch together a unified scale of magnitudes for the entire sky.[29] Herschel remained a consistent voice advocating study of variable stars, providing updated lists of known variable stars and their periods in his 1849 expansion of the *Treatise on Astronomy* and on through multiple editions.[30]

Herschel's speculation on *extrinsic* factors for η Argus' variability (obscuring clouds rather than stellar eruptions) was related to considerations regarding life in the universe that he had inherited from his father. William believed strongly that the universe was full of intelligent life, assuming that besides the moon and planets even the stars and sun were habitable. John Herschel was more circumspect in stating his own beliefs but nonetheless assumed stars hosted inhabited planets. In correspondence, he resisted the writings of his friend William Whewell (1794–1866), Master of Trinity College, Cambridge, who argued for a lifeless solar system beyond Earth. When it came to stars like η Argus, Herschel was concerned what such fluctuations meant for the inhabitants of its assumed planetary system.[31]

Solar System Nomenclature

Apart from visible observations of stellar magnitudes, after his return to England from South Africa, Herschel's observing days were over. The next decade was spent reducing his observations and preparing the *Cape Results* for publication. This meant that, when Neptune was discovered in 1846, Herschel was still working with data he had gathered from the Cape and in particular calculating the orbits of Saturn's moons, which had been well-positioned for observing during his time in South Africa. Herschel's response to controversies that arose over the discovery and naming of Neptune led, through the *Cape Results*, to an enduring nomenclatural system for the solar system still followed today. Prior to the discovery of Neptune, moons had not been given proper names but were instead only numbered. By introducing classical names for Saturn's moons, Herschel used his influence to circumvent national debates over naming objects while simultaneously (and perhaps inadvertently) leading to a naming system for the moons of Uranus that enshrined English literature around the planet originally named for an English king.

Though William had named the planet he discovered in 1781 the Georgium Sidus, John Herschel followed European convention and called it Uranus. The question of naming planets reemerged with Neptune's discovery based on the predictions of French astronomer and mathematician Urbain Le Verrier (1811–77). Le Verrier proposed his own name for the planet, and tempers flared as British astronomers claimed co-discovery credit for John Couch Adams (1819–92), a British mathematician and astronomer who had been pursuing similar, unpublished predictions, and on this basis proposed alternate names for the planet. In the midst of this controversy, which Herschel inadvertently stoked by his support of Adams' claim, Le Verrier wrote to Herschel as the only other astronomer with a personal stake in planetary naming to pledge he would henceforth refer to Uranus as *Herschel* if Herschel would support calling Neptune *Le Verrier*.[32]

Le Verrier's letter found Herschel in the midst of preparing his *Cape Results* chapter on the satellites of Saturn. At the time, Saturn was known to have seven satellites, two of which had been discovered by William in 1789.[33] William had not named the moons. Despite the early suggestion of Simon Marius (1573–1625) that the moons of Jupiter be called Io, Europa, Ganymede, and Calisto, satellites of the outer solar system had since their discovery been only numbered. This convention was complicated by Saturn's moons, however, as the order in which the moons were discovered did not match their order from the planet. The conventional designation for Saturn's moons at the time of Neptune's discovery was 6, 7, 1, 2, 3, 4, and 5, numbering outward from the planet. Despite this confused system, however, John Herschel continued to use it in his notes and correspondence.[34]

This changed the day Herschel received Le Verrier's letter regarding planet names. On that date, mythological names for the moons of Saturn make an appearance in Herschel's diary for the first time. When the *Cape Results* went to press shortly thereafter, the moons of Saturn bore mythological names in place of numbers: Mimas, Enceladus, Tethys, Dione, Rhea, Titan, and Iapetus. In a footnote, Herschel explained he had chosen the names to replace the confusing numbering system.[35] Since Herschel had never expressed frustration with the numbering system before this, however, and had used it in his correspondence up to this point, it is more likely he used the opportunity of establishing a new naming scheme to emphasize the suitability of mythological rather than contemporary, personal names in the solar system. The *Cape Results*, distributed to astronomers and observatories around the world, propagated this naming system, and when William Lassell (1799–1880) discovered an additional moon of Saturn in 1848, he dubbed it Hyperion in keeping with Herschel's new convention.

Herschel's influence led Lassell to approach him again when Lassell later confirmed four of William Herschel's supposed six moons of

Uranus. Lassell, considering William's priority of discovery, wanted Herschel to provide names for these moons as well. Herschel responded by suggesting names of sprites or "airy spirits" and offered Ariel, Umbriel, Titania, and Oberon.[36] None of these names, however, were from mythology. Instead, they all came from the works of English authors: Umbriel from Alexander Pope's *The Rape of the Lock* and the other three from Shakespeare's plays *The Tempest* and *A Midsummer Night's Dream*. Such was Herschel's influence that these were quickly accepted as well, though the convention quickly evolved away from fairies to exclusively Shakespearian characters. Ironically then, Herschel's attempt to de-nationalize the heavens resulted in the planet his father first named for an English king remaining surrounded by names from English literature.[37]

The General Catalogs

After Herschel's publication of the *Cape Results*, he continued corresponding with astronomers and gathering information for new editions of his *Outlines of Astronomy*, even while his own work became focused on research into photography, championing government sponsorship of science, involvement with the newly formed British Association for the Advancement of Science, and eventual responsibilities as Master of the Mint. It was not until the final decade of his life that he returned to the systematic catalogs for which he had gained such fame. In these, he still pursued utility for other observers and the establishment of empirical baselines for future observations. He began with the nebulae, having received striking reports from observers that some of his father's had significantly altered or even disappeared. As there were growing numbers of observers who now had telescopes that put these objects within reach, Herschel felt the need to organize and synthesize all observations made of sidereal objects since his earlier work and to consolidate his and his father's previous catalogs into a unified list. This "General Catalogue," presented to the Royal Society in 1864, the same year William Huggins (1824–1910) published his first spectroscopic analysis of nebulae, became the basis and model for the New General Catalogue, by which thousands of deep sky objects are known by their NGC numbers today.[38]

After this catalog, Herschel began the same process for double stars, ending his astronomical career where it began, collecting and summarizing the locations and data for all known double stars. In some ways more dramatically than the nebulae, double stars had transformed popular perceptions of the physical nature of stars during Herschel's lifetime, as observations and mathematical analysis showed them to be massive, dynamic objects – moving them in the rhetoric of Herschel's popular writings and the writings of other popularizers of astronomy from the sublime *landscape* of the heavens to be considered themselves sublime *objects*, sweeping out orbits at velocities so immense as to be visible from Earth. In the years leading up to the advent of astrophysics, double stars were a primary way in which stars gained physicality. This final project, however, Herschel would not live to see complete. He died in 1871, in the midst of compiling the ten thousand double stars of this, his last catalog.

If William and Caroline Herschel are seen as ushering in a sidereal revolution by turning the attention of observers beyond the solar system, it took the astronomical career of John Herschel to establish this revolution. Herschel used his mathematical training, scientific credibility, and social capital to move the study of nebulae, double stars, and variable stars from the sidereal fringes of astronomy to become an accepted part of observational practice. He did this through his own observations and catalogs as well as his personal connections and constant appeals to the astronomical community and wider public, showing through his work and example that these objects could lead to important physical insights of the sidereal universe. And, as the history of stellar astronomy has shown, he was right.

Notes

1 David Aubin, Charlotte Bigg, and H. Otto Sibum, "Introduction: Observatory Techniques in Nineteenth-Century Science and Society," in *The Heavens on Earth: Observatories and Astronomy in Nineteenth-Century Science and Culture* (Durham, NC: Duke University Press, 2010), 1–32.
2 David W. Dewhirst, "Meridian Astronomy in Private and University Observatories of the United Kingdom: Rise and Fall," *Vistas in Astronomy*, 28 (1985): 147–58, on 150. The article to which the reader is directed can be found in Richard Sheepshanks, "Transit Instrument," *The Penny Cyclopedia*

of the Society for the Diffusion of Knowledge (London: Charles Knight & Co, 1833–43), vol. 25, 122–34.

3 John Herschel, "An Address Delivered . . . on the Occasion of the Distribution of the Honorary Medals . . . on April 11, 1827, to Francis Baily, Esq., Lieutenant W. S. Stratford, R.N., and Colonel Mark Beaufoy," *Philosophical Magazine*, 2 (1827): 456–57.

4 William Herschel, "Catalogue of One Thousand New Nebulae and Clusters of Stars," "Catalogue of a Second Thousand New Nebulae and Clusters of Stars," and "Catalogue of 500 New Nebulae, Nebulous Stars, Planetary Nebulae, and Clusters of Stars," *Philosophical Transactions of the Royal Society*, 76 (1786): 457–99; 79 (1789): 212–55; and 92 (1802): 477–528. For an overview of William Herschel's contributions to astronomy see Michael Hoskin, *The Construction of the Heavens: William Herschel's Cosmology* (Cambridge: Cambridge University Press, 2012).

5 William Herschel, "Account of the Changes That Have Happened, during the Last Twenty-five Years, in the Relative Situation of Double-stars; with an Investigation of the Cause to which They Are Owing," *Philosophical Transactions of the Royal Society*, 93 (1803): 339–82.

6 Emily Jane Winterburn, "The Herschels: A Scientific Family in Training" (PhD diss., Imperial College London, 2011), 85.

7 William Herschel, "Observations of a Comet, Made with a View to Investigate Its Magnitude and the Nature of Its Illumination. To Which Is Added, an Account of a New Irregularity Lately Perceived in the Apparent Figure of the Planet Saturn," *Philosophical Transactions of the Royal Society*, 98 (1808): 145–63, on 160.

8 Philip C. Enros, "The Analytical Society (1812–1813): Precursor of the Renewal of Cambridge Mathematics," *Historia Mathematica*, 10 (1983): 24–47, but see also my *Creatures of Reason: John Herschel and the Invention of Science* (Pittsburgh: University of Pittsburgh Press, forthcoming).

9 See, for instance, William Ashworth, "The Calculating Eye: Baily, Herschel, and the Business of Astronomy," *British Journal for the History of Science*, 27.4 (1994): 409–41.

10 William Herschel, "Catalogue of Double Stars," and "Catalogue of Double Stars," *Philosophical Transactions of the Royal Society*, 72 (1782): 112–63 and 75 (1785): 40–126.

11 John Michell, "An Inquiry into the Possible Parallax and Magnitude of the Fixed Stars, from the Quantity of Light which They Afford to Us, and the Particular Circumstances of Their Situation," *Philosophical Transactions of the Royal Society*, 57 (1767): 234–64.

12 John Herschel to Caroline Herschel, 22 May 1831, Herschel Family Papers, Subseries C, Correspondence (henceforth TxU:H) L-0576.3, Harry Ransom Center, University of Texas, Austin.

13 John Herschel and James South, *Observations of the Apparent Distances and Positions of 380 Double and Triple Stars, Made in the Years 1821, 1822, and 1823, and Compared with Those of Other Astronomers; Together with an Account of Such Changes as Appear to Have Taken Place in Them since Their First Discovery* (London: Nicol, 1825).

14 John Herschel, "On the Parallax of the Fixed Stars," *Philosophical Transactions of the Royal Society* (1826): 266–80.

15 John Herschel, "On the Investigation of the Orbits of Revolving Double Stars . . .," *Memoirs of the Royal Astronomical Society*, 5 (1833): 171–222. See also Thomas L. Hankins, "A 'Large and Graceful Sinuosity': John Heschel's Graphical Method," *Isis*, 97.4 (2006): 605–33.

16 John Herschel, "Account of Some Observations Made with a 20-feet Reflecting Telescope," "Approximate Places and Descriptions of 295 New Double and Triple Stars . . .," and "Observations with a 20-feet Reflecting Telescope – Third Series; Containing a Catalogue of 384 New Double and Multiple Stars . . .," *Memoirs of the Astronomical Society of London*, 2 (1826): 459–97; 3 (1829): 47–63; and 3 (1829): 177–213.

17 John Herschel, "Micrometrical Measures of 364 Double Stars with a 7-feet Achromatic Telescope, Taken at Slough, in the Years 1828, 1829, and 1830," *Memoirs of the Royal Astronomical Society*, 5 (1833): 13–91.

18 R. Main and C. Pritchard, eds., "A Catalogue of 10,300 Multiple and Double Stars Arranged in the Order of Right Ascension by the Late Sir J. F. W. Herschel, Bart.," *Memoirs of the Royal Astronomical Society*, 40 (1874): 1–144.

19 Wolfgang Steinicke, *Observing and Cataloguing Nebulae and Star Clusters: From Herschel to Dreyer's New General Catalogue* (Cambridge: Cambridge University Press, 2010), 38–41.

20 John Herschel, "Observations of Nebulae and Clusters of Stars, Made at Slough, with a Twenty-feet Reflector between the Years 1825 and 1833," *Philosophical Transactions of the Royal Society*, 123 (1833): 359–505, on 504.

21 George Biddell Airy, "President's Address upon Awarding the Medal to Sir John J. F. W. Herschel," *Monthly Notices of the Royal Astronomical Society*, 3 (1836): 167–74, on 170, 173–74.

22 Gregory Good, "J. F. W. Herschel's Optical Researches: A Study in Method" (PhD diss., University of Toronto, 1982), on 38.

23 For the details of Herschel's dispute on this topic, see Stephen Case, *Making Stars Physical: The Astronomy of Sir John Herschel* (Pittsburgh: University of Pittsburgh Press, 2018), 160–66.

24 John Herschel, *Results of Astronomical Observations Made during the Years 1834, 5, 6, 7, 8, at the Cape of Good Hope; Being the Completion of a Telescopic Survey of the Whole Surface of the Visible Heavens, Commenced in 1825* (London: Smither, Elder, and Co., 1847).

25 John Herschel to Thomas Maclear, 16 December 1837, printed in Brian Warner and Nancy Warner (eds.), *Maclear & Herschel: Letters & Diaries at the Cape of Good Hope, 1834–1838*. (Cape Town: A. A. Balkema, 1984), 207–8.

26 John Herschel to Heinrich Christian Schumacher, 19 January 1839, published in *Astronomische Nachricheten*, 16 (1839): 187–90.

27 Herschel, *Cape Results*, §273, §275.

28 John Herschel, "On the Suspected Variability of the Star α Cassiopeia," and "On the Variability and Periodical Nature of the Star α Orionis," *Monthly Notices of the Royal Astronomical Society*, 4 (1839): 215–16 and 11 (1840): 269–78.

29 John Herschel, ed., *A Manual of Scientific Enquiry; Prepared for the Use of Officers in Her Majesty's Navy; and Travelers in General*, 3rd ed. (London: Murray, 1859).

30 John Herschel, *Outlines of Astronomy* (London: Longman, Brown, Green, & Longmans, 1849), §819, §827.

31 Michael J. Crowe, *The Extraterrestrial Life Debate, 1750–1900* (Mineola: Dover, 1999), 216–21, 310–11. Herschel, *Cape Results*, §78.

32 Urbain J. J. Leverrier [*sic*] to John Herschel, 28 November 1846, Herschel Papers, Royal Society, 11.197.

33 William Herschel, "Account of the Discovery of a Sixth and Seventh Satellite of the Planet Saturn; with Remarks on the Construction of Its Ring, Its Atmosphere, Its Rotation on an Axis, and Its Spheroidal Figure," *Philosophical Transactions of the Royal Society*, 80 (1790): 1–20.

34 For details, see Stephen Case, "A 'Confounded Scrape': John Herschel, Neptune, and Naming the Satellites of the Outer Solar System," *Journal for the History of Astronomy*, 50 (2019): 306–25.

35 Herschel, *Cape Results*, footnote to §398.

36 William Lassell to John Herschel, 3 November 1851, Herschel Papers, Royal Society, 11.141, and John Herschel to William Lassell, Herschel Papers, Royal Society, 25.14.18.

37 Todd A. Borlik, "Stellifying Shakespeare: Celestial Imperialism and the Advent of Universal Genius," *Shakespeare in Southern Africa*, 26 (2014): 1–12.

38 John Herschel, "Catalogue of Nebulae and Clusters of Stars," *Philosophical Transactions of the Royal Society*, 154 (1864): 1–137.

4

Stargazer at World's End
John Herschel at the Cape, 1833–1838

From 1833 to 1838, John Herschel relocated himself, his growing family, and one of his father's largest telescopes to the British colony of Cape Town, in South Africa, with the intent of undertaking the most thorough observational study of the southern heavens ever made. This endeavor was driven in part by his desire to complete the astronomical work begun by his father William, whose own astronomical surveys had been confined to the northern hemisphere. The southern heavens had been previously mapped by Edmond Halley (1656–1742) and the Abbé Nicolas-Louis de Lacaille (1713–62), but their surveys were primarily undertaken for navigational purposes and with comparatively small telescopes.

To prepare for this immense project, in 1825 John set about resurveying the northern night sky, searching in particular for new nebulae, star clusters, and double stars, which had been William Herschel's primary objects of interest. These surveys gave Herschel the opportunity to practice using his father's twenty-foot telescope and to revisit objects his father had discovered and provide more accurate positions and more detailed descriptions (see Chapter 3). As a result of these surveys, Herschel discovered over five hundred new nebulae and star clusters, publishing a catalog of the work (totaling some 2,306 objects) in 1833. For this he received the Gold Medal of the Royal Astronomical Society and the Royal Medal of the Royal Society.

Herschel knew that a proper survey of the southern sky required significantly more time and a much larger telescope than what Halley and Lacaille had employed. He already had the bigger telescope and by the early 1830s he had the time. In 1830, he lost the election for president of the Royal Society to Augustus Frederick (1773–1843), the Duke of Sussex and brother of King George IV. This loss was actually a relief, as it saved John from the tedious duties the office would bring. Two year later, his mother, Mary Pitt Herschel (1750–1832), passed away. With these personal ties gone and having stepped down as president of the Royal Astronomical Society, Herschel had the freedom to consider a voyage to the southern hemisphere. He had a considerable fortune that came in part from the money his father made selling telescopes as well as from the estate he inherited from his mother, who had been a widow before she married William. In short, he now had the funds needed to undertake the voyage in the manner he preferred – at his own expense, "responsible to no one."[1] Such independence was essential, given his vision for how his observations were to be conducted.

An Unofficial Voyage

After careful consideration, Herschel chose Cape Town as the ideal location for an observatory. Why the Cape? One reason was that, although located at the southern tip of Africa, the Cape shared the same longitude as Western Europe. This would enable observations made there to be more easily correlated with observations in Europe. The climate was mild, which would help ensure clear skies. In addition, the Cape colony had been under British possession since 1806 and was a relatively stable outpost of the empire. This was important, as John would be bringing his wife Margaret and their growing family with him. Finally, it had its own Royal Observatory, established in 1820 and funded by the Admiralty to extend the work of the Royal Observatory at Greenwich to the southern hemisphere. Although Herschel did not intend to be officially tied to the Royal Observatory at the Cape, he would benefit from its presence.

Herschel did not want his voyage to have any official ties to the British government. The Admiralty, however, did not see things the same way. Following the defeat of Napoleon in 1815, the British Admiralty began a new phase of state-sponsored exploration, especially in Africa. John Barrow (1804–45), the Second Secretary of the

Admiralty and cofounder of the Royal Geographic Society, was fascinated with empty spaces on maps. After 1815, with the Royal Navy largely idle, he used his position to attempt to fill in some of those blank spaces. In 1816 he wrote, "To what purpose could a portion of our naval force be ... especially in a time of profound peace, more honourably or more usefully employed than in completing those details of geographical and hydrographical science of which the grand outlines have been boldly and broadly sketched by Cook ... and other of our countrymen?"[2]

An example of the sort of expedition that Barrow hoped the navy could support was that of Richard Lander (1804–34) and John Lander (1806–39) to the headwaters of the Niger River in 1830. When the Lander brothers returned to England in 1831, they immediately became national heroes. The *Edinburgh Review* called their expedition "the most important geographical discovery of the present age" and "a just ground for national glory."[3] The Landers' published journal of their travels became a widely popular work and brought new life to the goal of African exploration. It is significant then, that, at about the same time the Landers' discovery caused such a sensation and provided new promise for British commerce and control within Africa, John Herschel made public his own intended African voyage. Any expedition to Africa after 1831 invariably provoked national interest, because it invoked national pride.

By mid-1832, as Herschel prepared to depart for the Cape, he received his first offers of assistance from the Admiralty. From the outset, however, he intended his voyage to be a "private adventure." It would be a family adventure as well, including his wife Margaret (1810–84) and their first three children: Caroline (1830–1909), Isabella (1832–93), and William (1833–1917). Also accompanying the Herschels was John's mechanic John Stone, a servant James Rance, and their children's nurse Mrs. Nanson, who died soon after their arrival in Cape Town in 1834. Not only would private passage to and from the Cape for this family entourage be expensive, but there was also the matter of transporting his large twenty-foot telescope and other instruments. Furthermore, Herschel would purchase a house and property for the duration of their stay. Despite these expenses, Herschel was prepared to pay whatever was necessary so that he might survey the southern heavens on his own terms.

Herschel was given the opportunity of having much (and later all) of his proposed voyage paid for by the British government. But he turned down these offers – repeatedly. A likely explanation is Herschel's support of his friend Charles Babbage's complaint, in his *Reflections on the Decline of Science in England, and on Some of its Causes* (1830), that lack of government support for the sciences meant English science would continue falling behind that of France and Germany. Even before Babbage's book, Herschel had refused his final stipend for serving as a commissioner on the Board of Longitude after the Board was disbanded by the government in 1828 in what Herschel believed was an egregious lack of support for British men of science by the Admiralty. If the government could not be bothered to support the sciences, it would be a "foul shame" for him to "pocket the wages of a Government which treats its agents so cavalierly."[4]

Unsurprisingly then, a few years later, he also rejected the first offer of support for his trip to the Cape, possibly thinking that in light of his previous stance it would be hypocritical to do otherwise. The offer came from the president of the Royal Society, Augustus Frederick (to whom Herschel had narrowly lost the election), who proposed free passage on a vessel of the Royal Navy. Herschel politely declined, downplaying the scientific intent of his voyage to make it appear more like a family vacation:

> With regard to ... passage in one of the Government vessels ... I am perfectly aware of the superior comfort and advantages of that mode of conveyance, over what would be afforded by a private or Company's ship but ... I confess I do not see how, consistently, with the view I entertain of the project, which is that of an entirely irresponsible private adventure (a mere party of pleasure in short) I could avail myself of any application to the Admiralty ... the effect of which would be not only to confer something of an official character to the undertaking, but to shift from my own shoulders on the Public a considerable ... expense which I have resolved to meet, as a necessary attendant on the execution of it.[5]

In other words, Herschel wanted the independence of not feeling that the results of his labors were in some respect the property of the government or even the public.

In September, the Admiralty tried again, offering transportation in any form Herschel would accept, "managed so as to suit Sir John's conveniences," especially in case he had problems transporting his telescope and other equipment on a private ship.[6] Herschel insisted he would go privately or not at all:

> *Should* such occur and prove insuperable I shall though with extreme reluctance abandon the project, rather than take any step which may give it the slightest tincture of an official character . . . . As to the expense of the voyage – of course it will be heavy, but having all along included it in my estimates, as a necessary item it ceases to be thought of as a hardship or to excite any desire to be relieved from it by a free passage in the King's ship.[7]

Still, the offers continued. Sir James Graham (1792–1861) of the Naval Board wrote to Herschel's friend Basil Hall that he considered "the voyage of S[ir] John Herschel to the Southern Hemisphere so intimately connected with the highest objects of science & with the Naval Service, that I am anxious to give him every assistance, & should be most happy . . . for ordering him & his attendants a passage in a King's ship to the Cape of Good Hope." He urged for John to "wait a little, & see . . . before he makes any private arrangement for his Voyage, I still may be able to serve both him & the Public by offering him some facility." Hall wrote Herschel that Graham's message "almost amounts to an order for me to go to Slough [Herschel's home] with a party of Marines, & force you at the point of the bayonet to accept a passage in a man of war!"[8]

Still Herschel refused, and when a fourth offer was made (again by Augustus Frederick) in May of 1833, he made it clear he was not interested in passage on a government vessel under any circumstances, with but one exception:

> There is only one contingency which would induce me to look on the matter differently – viz: a declaration of war with a maritime power – In such an event, should the voyage become admittedly unsafe – the security of such a ship might be very enviable – but on the other hand in that event the King's ships would have other fish to fry than landing stargazers at the world's end.[9]

Having stood his ground against well-wishing supporters who, he felt, would alter the tenor of the work he was conducting and make both him and his result beholden to government and public interest, Herschel and his family left England on November 13, 1833, aboard a private ship, the *Mountstuart Elphinstone*, a 611-ton, Bombay-built vessel that transported both passengers and cargo. After a two-month voyage, they arrived at the Cape without incident on January 16, 1834. The following March, Herschel could write to his aunt, Caroline Herschel (1750–1848) that, "We have been very attentively received by almost all the resident English & Dutch families of the best note, so that if we wish it we may form a very large & good acquaintorium [*sic*] but we shall not enter into very much company."[10] Although the Herschels did not participate extensively in the social life of the colony, they would sometimes dine with other families, on which occasions John entertained locals eager to have a peek through his telescopes.

Other than his wife Margaret, Herschel had but one close confidant during his stay at the Cape: Thomas Maclear (1794–1879), the newly appointed astronomer at the Royal Observatory at the Cape. The two astronomers arrived in Cape Town within ten days of each other and became fast friends, corresponding almost daily, traveling together through the colony, and aiding each other in their scientific projects, which included Maclear's duties of regular magnetic and meteorological observations. A major difference between the two was that while Maclear had an observatory waiting for him upon his arrival, Herschel had to establish his own.

Feldhausen

Soon after he arrived in South Africa, Herschel purchased an estate about six miles outside of Cape Town. It was called Feldhuisen by its Dutch owner and the Grove by the English residents of the colony. The Herschels opted for a compromise and, drawing upon Herschel's German ancestry, renamed the estate Feldhausen. The house was in a beautiful setting, lush and isolated, with Table Mountain looming in the distance. A few years later, in 1836, the young Charles Darwin (1809–82) visited Herschel when the *Beagle* stopped at the Cape. He described Feldhausen as "a very comfortable country house, surrounded by fir and oak trees, which alone in so open a country, give a most charming air of seclusion and comfort."[11] The Herschels took possession of Feldhausen on April 23, 1834. Almost immediately, Herschel, his mechanic John Stone, and four local laborers set up his observatory on the estate's grounds.[12]

Figure 4.1 Twenty-foot telescope at Feldhausen

Herschel brought three telescopes with him to the Cape: his father's twenty-foot reflecting telescope (Figure 4.1) of 18.25-inch diameter, a seven-foot equatorial-mount refracting telescope of 5-inch diameter, and a reflecting "comet sweeper" of 9-inch diameter. To support the twenty-foot telescope, he brought the massive A-frame scaffolding that had held it in England. This secured the telescope in the middle, along with providing a platform above the instrument on which Herschel stood to look into the eyepiece, which focused the light reflected up from the mirror at the bottom of the tube. The platform and telescope were raised and lowered by a system of pulleys, and the entire structure rotated by means of wheels set on a circular track. While Herschel observed, his mechanic adjusted the telescope. On February 22, the twenty-foot telescope was assembled, and that night the first object Herschel observed was the star α-Crucis in the constellation of the Southern Cross.[13]

Nights spent observing could be long and tedious. For hours on end, Herschel and Stone worked together in near silence. Before he left for the Cape, astronomical observations occasionally brought Herschel to despair, as he wrote to his wife on one occasion: "Two stars last night and sat up until two waiting for them. Ditto the night before. Sick of star-gazing – mean to break the telescopes and melt the mirrors."[14] But at the Cape, under new skies and in Feldhausen's bucolic setting, the experience of observing became supremely gratifying. Soon after his arrival, Herschel gushed to his aunt Caroline, describing the southern heavens as "rich in stars, nebulae, and clusters beyond anything you can imagine."[15] Indeed, after years in the north, surveying objects his father had already charted before him, it is not hard to imagine Herschel's excitement at turning his telescope on a new celestial frontier. On the night of February 5, 1835, for instance, he recorded, "Made a long nights Sweep, and the night being most superb – the mirror brilliant and the zone swept . . . the richest perhaps in the heavens – attained the sublime of Astronomy – a sort of *ne plus ultra* . . . . [I]t is an epoch in my Astron^l life."[16]

As he and his father had done in the northern hemisphere, Herschel surveyed the southern heavens in "sweeps" at a rate of about one sweep per night when the skies cooperated. A sweep was a horizontal survey of a band of the night sky three degrees in width. During a sweep, the twenty-foot telescope remained pointed in the same direction, only moving up and down to scan the width of three degrees. Small bells marked the upper and lower limits of the sweep; they would chime when the telescope reached its maximum or minimum altitude. Lateral motion came from the rotation of the earth itself, which brought the stars, nebulae, and clusters into view of the telescope. Herschel gave a more complete description of the process:

> [A] handle is constantly kept moving to and fro, by which the telescope is kept oscillating up and down and "sweeping" over an arc in Polar distances usually limited to 3°. On the entry of any object into the field the motion is arrested by an order to the assistant and by directing him, up and down, slow or fast, much or little, the object is easily brought to a bisection on a horizontal wire, to any degree of nicety. To warn the attendant when to reverse the motion of the handle a bell is made to strike at each limit of the sweep, and these bells differ in pitch, to indicate which limit is attained, the higher pitch corresponding to the top and the lower to the bottom of the zone.[17]

By the end of 1835, Herschel's sweeps for nebulae, clusters, and double stars were nearly complete. He had cataloged 1,708 nebulae and star clusters (1,268 of which had never before been recorded) and 2,102 pairs of double stars. These objects, however, were not his only astronomical interests. Like his father, Herschel also made a series of star-gauges, counting the number of stars visible in roughly 2,300 sections of the sky (each section being ten minutes right ascension by three degrees polar distance). In this manner, he tallied a total of 68,948 stars.[18] In October 1835, Herschel observed the return of Halley's comet. He and Maclear studied the comet for the next seven months from both Feldhausen and the Cape Royal Observatory. From 1835 to 1837, Herschel also observed the satellites of Saturn. He was able to locate Mimas and Enceladus, two of Saturn's inner moons, which had not been seen since his father had first detected them in 1789. He made regular observations of sunspots, as well as two of the largest and most spectacular objects in the southern heavens, the Magellanic Clouds. Many in Europe awaited the results of Herschel's observations in the southern hemisphere primarily for his description of these two spectacular objects.[19]

Herschel also took measurements with his actinometer and astrometer. The actinometer was a device of Herschel's own invention, designed to measure solar radiation. Composed of a cylindrical vessel filled with water darkened with ink, it was exposed to sunlight at noon for a few minutes. Afterward, the temperature of the liquid was measured with a highly sensitive thermometer, with the goal of gauging how much energy the liquid had absorbed from sunlight.[20] Unfortunately, Herschel's many actinometer measurements were discarded after his returned to England when he realized he had not consistently used the same volume, or formula, for the liquid. This resulted in a number of anomalies in his data, and he was forced to throw out over four hundred of his observations.[21] (The astrometer, a device for measuring and comparing the intensity of starlight, was another invention of Herschel's, of which more below.)

In addition to his astronomical work, Herschel helped reform the colony's education system, which had to that point been "poorly administered and generally ineffectual," by lending his reputation and influence to help turn it into one of the "best structured state education systems in the world" (see Chapter 1).[22] This included working closely with the Colonial Secretary at the Cape, Sir John Bell (1782–1876), and others both at the Cape and following his return to England. Herschel advocated for government support of Cape schools to include funding for facilities, a standardized syllabus, and teachers recruited from England. Within a few months of his 1838 return to London, he petitioned Charles Grant, Lord Glenelg (1778–1866), Secretary of State for the Colonies, and secured £3,460 for the Cape's new educational scheme – one that, as Herschel and his colleagues intended, should admit students "without the least reference to colour or class."[23]

Ever curious about the flora and fauna around him, Herschel explored the area around Cape Town hunting, gathering specimens, and – with the assistance of his camera lucida and his artistically talented wife – drew superb images of flowers and plants of the Cape biota.[24] Of course, his scientific mind ranged over other pursuits, such as investigating the Cape's geology and making meteorological and tidal observations. These last were undertaken for his friend William Whewell (1794–1866) who, starting in 1833, commenced a global study of tides, relying on a network of observers all over the world to provide data. Herschel also occasionally invited the colony's residents to join him at his telescope. These visits, which Herschel referred to as "tea and stars," became especially frequent during the appearance of Halley's comet. Thomas Maclear reflected on the Herschel's hospitality at Feldhausen, writing years later that "the occasional gatherings at 'Feldhausen' consisted of the élite of the Cape."[25]

By the time he departed Africa, Herschel had made a significant impact on the social, cultural, and scientific life of the colony. The Cape also made an impact on him, including his approach to his astronomical surveys. Historian Elizabeth Green Musselman has described how Herschel fancied himself a "hunter" and "bagger" of stars when he arrived at the Cape. However, the shocking "mutual slaughter" of the Sixth Frontier War (1835) between the native Xhosa and the Colony's white settlers to the east of Cape Town, and the "dissolution of slave society into vagrancy," compelled him to realize "the need for maturity in his astronomical work." Subsequently, in Herschel's notes and diaries, Green Musselman suggests his language of *hunting* stars switched to an idiom of *cultivation*. For example, on February 7, 1837, he began a written register of the nebulae discovered during his sweeps by noting in his diary, "this is the beginning of my Cape *harvest*." At least in his

Figure 4.2 Herschel memorial obelisk at Feldhausen

metaphors, the violence Herschel witnessed at the extremities of the British empire tempered his language.[26]

Herschel sold Feldhausen when he returned to England, with the provision that the area where the twenty-foot telescope had stood would remain in his possession indefinitely. Soon afterward, the residents of the Cape erected a stone obelisk there (Figure 4.2). It was a monument not only to Herschel but also to the colony itself, which for over four years played host to "the most justly celebrated astronomer and philosopher of modern times."[27]

The Great Moon Hoax

Herschel's most famous (or perhaps, infamous) discovery while at the Cape was not, in fact, a real discovery at all. In August of 1835, far away from Cape Town, Herschel was an unknowing participant in a fantastic "hoax." On August 25, the *Sun*, a small newspaper in New York City ran a front page story with the headline "GREAT ASTRONOMICAL DISCOVERIES Lately Made By Sir John Herschel, LL D., F.R.S., &c, At the Cape of Good Hope."[28] The story went on to claim that it had learned from the most recent issue of the *Edinburgh Journal of Science* (a real publication but one that had gone out of print by the time the *Sun* "referenced" it) that Herschel had discovered life on the moon. Over the course of a few days, the *Sun* reported that Herschel had found (with the aid of a fictitiously large telescope) an amazing variety of plant and animal life on our satellite. After four days, the story culminated with the revelation that Herschel had observed rational beings on the moon: humanoid creatures with bat-like wings who demonstrated an ability to socialize and communicate.

As a result, the *Sun*'s circulation briefly became greater than that of any other American paper. Every available copy was snatched up in a frenzy of excitement. The reading public seemed unanimously convinced of the authenticity of the account. One clergyman even told his parishioners he would consider taking up a collection to send Bibles to the lunar inhabitants.[29] This was not, however, mass delusion. For many Americans and Europeans at the time, the existence of extraterrestrial life was considered highly probable, if not certain. Intelligent beings were assumed to exist not only on the other planets in our solar system but also on our moon and perhaps even the sun. The celebrated author Thomas Paine, for instance, in his *Age of Reason* assumed the existence of extraterrestrial life as part of his argument against Christianity.[30] Herschel's own belief that the planets of the solar system were habitable was clearly spelled out in his *Treatise on Astronomy*, a work that enjoyed considerable popularity in America. It therefore hardly seemed surprising to the readers of the *Sun* that life had been found on the moon – or that it was John Herschel who discovered it.

Soon after the last installment on August 31, the story was revealed to be satire crafted to demonstrate the gullibility of Americans on the issue of extraterrestrial life (and to sell newspapers). The author of the story, Richard Adams Locke (1800-71), had effectively mocked the assumption long held by many that intelligent life was certain to exist on other planets. The fact that nearly everyone in the major American cities believed the piece has wrongly elevated the whole episode to the level of hoax, and it is commonly remembered as "The Great Moon Hoax." But, as Locke himself wrote, "it is quite evident that it is an abortive satire." Yet in its success as news, it failed as satire; everyone had been fooled. Edgar Allen Poe recalled, "Not one person in ten discredited it . . .. A grave professor of mathematics in a Virginia college told me seriously that he had no doubt of the truth of the whole affair!"[31]

From Locke's perspective, Herschel was one of the few astronomers whose name could give credibility to his satire. And, because Herschel was in Africa at the time, it was impossible for the astronomer to contradict it. Besides, what more appropriate place than the farthest point on the mysterious continent of Africa from which to discover extraterrestrial life? For many Americans in the 1830s, southern Africa was as remote and fantastical as a fairy tale. It may as well have been the moon. When Herschel first learned about Locke's satire, he had little to say. He was not pleased at the way he had been used, but mostly he regretted the flurry of letters he received from those who had either not heard that the story was a fabrication or simply wanted his opinion on the matter. He commented on the situation to his aunt Caroline: "I have been pestered from all quarters with that ridiculous hoax about the Moon – in English French Italian & German!!"[32] Years later, in 1842, the mathematician Augustus de Morgan (1806–71) informed Herschel that many people still believed in the truth of the Moon Hoax.[33]

Some stories were just too good to let go.

Return and Reception

Herschel and his family left the Cape and returned to England in 1838, at which point his career as an observational astronomer was effectively over. With but a few minor exceptions, he would never again use his telescopes. "With the publication of my South African observations (when it shall please God that shall happen)," Herschel proclaimed, "I have made up my mind to consider my astronomical career as terminated."[34] His return also represented an end to the privacy he so much enjoyed while at the Cape, years he would describe as full of "a thousand pleasing and grateful recollections" and "spent in agreeable society, cheerful occupation, and unalloyed happiness."[35] Once back in

London, he was swept up in the hectic life of the metropolis. Although the Herschels moved forty miles outside the city to a house called Collingwood, near Hawkhurst, Kent, his numerous scientific passions kept him actively involved with Britain's scientific circles, and he often maintained an apartment in London.

About a month after his return, on June 28, 1838, Herschel was made a baronet – a low-ranking but hereditary title. Appropriate to the esteem in which he was held, this honor was presented in Westminster Abbey at the coronation of Queen Victoria.[36] A few weeks before the coronation, on June 15, 1838, a public dinner was held in his honor. A full account of the dinner is given in the Saturday, June 16, 1838, edition of the *Athenaeum*. The list of attendees reads like a "who's-who" of early Victorian science. Held at the Freemason's Tavern in London, it included (to name a very few) the Irish mathematician William Rowan Hamilton (1805–65), the geologists Roderick Murchison (1792–1871) and Charles Lyell (1797–1875), the Scottish physician William Somerville (1800–78, husband of the mathematician and scientific writer, Mary Somerville), and Charles Darwin. Even the physicist Michael Faraday (1791–1867), who never attended public events, made an exception. (When, a few years later, Faraday declined an invitation to dine with Herschel, he recalled with wry humor, "Many thanks for your kind invitation which I should most gladly accept but that I never dine out on any occasion. I believe the last time I did do so was to dine with you at the Freemasons Hall & that was a solitary occasion.")[37] Herschel's closest scientific friends – William Whewell, Charles Babbage (1791–1871), and George Peacock (1791–1858) – were there, as were the Arctic and Antarctic explorers William Edward Parry (1790–1855), John Ross (1777–1856), and James Ross (1800–62), along with members of the Admiralty such as Herschel's friend and ally Basil Hall (1788–1844). The list also included lords, earls, bishops, members of Parliament, military officers, and professors. All in all, estimated the *Athenaeum*, there had been at least four hundred people present to honor Herschel's return.[38]

As this fête showed, upon his return from Africa Herschel had become imperial Britain's premiere man of science, a sort of *natural philosopher laureate*. Overwhelmed by the pomp that attended his homecoming, he was torn between expectation of his involvement in a variety of British scientific societies and projects and the simple, private family life he preferred. Furthermore, he still faced the daunting task of reducing his Cape observations before putting them into print. This would require extensive calculations for each and every object observed in the southern sky.

Production of the *Cape Results*

In addition to the calculations, there remained the task of organizing his observations into a form comprehensible to the scientific world and preparing the astronomical drawings he made for engraving. Furthermore, there was the theoretical matter of interpretation: some of the phenomena Herschel had observed had no clear scientific explanation. There were also his other personal scientific interests, long subordinated to his astronomical observations, which presented themselves anew. On top of all this, his return to England left him vulnerable once again to the demands that others – especially the scientific community – placed upon him. In the first years following his return, everything seemed to conspire against him working on his Cape observations. Nearly a year after his return from the Cape, he wrote to his friend James David Forbes (1809–68), "I have been entirely disabled from advancing a step in the reduction of my Cape observations, which remain precisely (with one most trifling exception) in the state in which they were when I left the Cape."[39]

Herschel spent nearly seven years reducing his Cape observations, finishing on March 7, 1847 (his fifty-fifth birthday). A generous grant from Hugh Percy (1785–1847), Duke of Northumberland, enabled the observations to be published in one large volume, as Herschel preferred, rather than piecemeal in journals. The work was published by Smith, Elder, and Company of Cornhill, London. (Herschel's brother-in-law, Peter Stewart was an employee of Smith, Elder, and Company and likely convinced Herschel to use them.) Final cost of publication came to just over one thousand pounds (the equivalent of £73,000 or over $80,000 today according to the Bank of England's online inflation calculator), which was almost entirely covered by Percy's grant. The book was in print by June of 1847, bearing the imposing title of *Results of Astronomical Observations Made During the Years 1834, 5, 6, 7, 8, at* the Cape of Good Hope; Being the Completion of a Telescopic Survey of the Whole Surface of the Visible Heavens, Commenced in 1825 but generally referred to as the Cape Results. It is a large quarto volume (approximately 12.5" x 10.5"), 452 pages in length (xx, plus appendices, errata, additions, and corrections), and has numerous illustrations at the back. The Cape Results is a largely empirical work. Most of the chapters are in effect astronomical catalogs: tables of data of positions and descriptions of celestial objects by an astronomer for other astronomers. The first four chapters, which comprise more than four-fifths of the book, consist almost entirely of such tables. The last three chapters are relatively short and are descriptive (chapter 6) or theoretical (chapters 5 and 7). The seven chapters, discussed briefly in turn, are:

1 Of the Nebulae of the Southern Hemisphere
2 Of the Double Stars of the Southern Hemisphere
3 Of Astrometry, or the Numerical Expression of the Apparent Magnitudes of the Stars
4 Of the Distribution of Stars, and of the Constitution of the Galaxy in the Southern Hemisphere
5 Observations of Halley's Comet, with Remarks on its Physical Condition, and that of Comets in General
6 Observations of the Satellites of Saturn
7 Observations of the Solar Spots

The first chapter is a lengthy 164 pages – one-third of the entire book. In it, Herschel provides descriptions and positions of the 1,708 nebulae and star clusters he observed in the southern hemisphere, along with his "law of distribution" of those objects. This chapter also includes detailed descriptions of the Magellanic Clouds. The second chapter, on double stars, complimented his father's study of those same objects in the northern hemisphere.[40] The emphasis in John Herschel's work is on their potential role in determining stellar distance via parallax. Chapter 3 considers Herschel's method for expressing stars' apparent magnitude with his astrometer. Without any way to obtain the absolute magnitude of stars, astronomers had to rely on visual methods of ordering the apparent brightness of stars relative to each other. Herschel's method involved reflecting moonlight through a prism, which was then focused with a lens so that the light would "appear ... as a star." With the astrometer, an observer could compare the relative brightness of stars using the "absolute" brightness of the moon as a baseline or standard. (Herschel even had an equation that took into account the moon's phase and thus its brightness, admitting later that it would have been easier to use a planet such as Jupiter as his standard.)[41]

In chapter 4, Herschel describes the Milky Way as seen from southern latitudes, with an emphasis on the varying density of the galaxy's stars, which he arrived at using careful gauges, or counts, of stars in regions of the Milky Way. The purpose of these gauges was to obtain statistical information on the distribution of stars. As a result of his northern hemisphere star gauges, William Herschel had been one of the first astronomers to propose a structure for our galaxy, proposing that the Milky Way was a "stratum of fixed stars."[42] John, however, during his own resurvey of the northern hemisphere and especially after his southern hemisphere star gauges, rejected the "stratum" structure, determining instead that the distribution of stars and nebulae in the Milky Way was far less uniform than his father suggested. Herschel described the structure of our galaxy as an "annulus," a ring-like structure, but one with "discontinuous masses and aggregates of stars in the manner of the cumuli of a mackerel sky, rather than a stratum of regular thickness and homogenous formation."[43]

In chapter 5, Herschel describes his months-long observations of Halley's comet, proposing a controversial theory on the nature of comets – in particular, why their heads and tails change size and direction as they approach and recede from the sun. In order to explain the way Halley's comet changed at perihelion, Herschel speculated that there was a secondary (and repulsive) force acting on it in addition to gravity: namely, electricity. However, in the mid-nineteenth century, this was tantamount to invoking an occult force, and Herschel received pushback by reviewers of the *Cape Results*. For example, David Brewster (1781–1868) lamented that by admitting "electricity ... in our sidereal systems, the mesmerists and phrenologists will form an alliance with the astrologer, and again desecrate with their sorceries those hallowed regions on which the wizard and conjuror have ceased to tread."[44] Historian Brian Warner, however, believes that Herschel had in fact provided the first "quantitative evidence for the action of non-gravitational forces in cometary phenomena."[45] In chapter 6,

Herschel describes his observations of the moons of Saturn. Finally, in chapter 7, Herschel hypothesizes on the nature of sunspots. He suggests that the existence of "solar spots" had to do with the sun having an atmosphere of its own, noting that he always observed the spots in mid-latitudes on the sun's surface, a region that correlated to the areas on Earth where "hurricanes and tornadoes prevail."[46]

Five hundred copies of the first and only edition of the *Cape Results* were printed. Three hundred fifty were given away by Herschel and Percy, including elaborate presentation copies distributed to royalty and heads of state across Europe and less elaborate copies to observatories around the world. The remaining one hundred fifty copies were allotted to the publisher for public sale. After publication, the *Cape Results* was reviewed by prominent men of science in a number of popular and academic publications, including the *Athenaeum*, the *Spectator*, the *Quarterly Review*, *Fraser's Magazine*, *North British Review*, and the *American Journal of Science and Arts*.[47] By and large, the praise was effusive, but there were criticisms as well. On the whole, however, it was lauded as a magnificent achievement, a legacy that took two generations of Britain's greatest astronomers to complete. In the words of David Brewster, reviewing the *Cape Results* in the *North British Review*:

> In the history of Astronomical Discovery there shine no brighter names than those of Sir William and Sir John Herschel – the father and the son. It is rare that the intellectual mantle of the parent lights upon the child ... [but in] the universe of mind, the phenomenon of a double star is more rare than its prototype in the firmament, and when it does appear we watch its phases and its mutations with a corresponding interest .... The records of Astronomy do not emblazon a more glorious day than that, in which the ... arc of the father was succeeded by the ... arc of the son. No sooner had the evening luminary disappeared amid the gorgeous magnificence of the west, than the morning star arose, bright and cloudless in its appointed course.[48]

Conclusion

The historian Walter (later Susan Faye) Cannon summarized the significance of Herschel's Cape voyage this way: "In November 1833

[Herschel] sailed for the Cape of Good Hope to observe the nebulae of the southern heavens, and sailed into apotheosis. After 1833 it was very difficult for an English writer to discuss, or even mention, Sir John Herschel in a scientifically impersonal tone."[49] Herschel, however, treasured his time in Cape Town not because of the fame it brought him but for highly personal reasons. The years spent with his family in the Cape colony were among the happiest of his long and productive life, a life filled with countless pleasant scientific *and* personal memories. Although he declined to serve as president of the Royal Society upon his return in 1838 (a position that was, the second time around, his for the taking), he did serve as Master of the Mint from 1850 to 1855. He also became involved the Admiralty's *Manual of Scientific Enquiry* and used his influence in support of the globe-spanning "Magnetic Crusade."[50] As a result of the Cape voyage and the publication of the *Cape Results*, Herschel became the living embodiment of Victorian science until his death in 1871, when he was interred in Westminster Abbey near the tomb of Sir Isaac Newton.

In the century and a half since his death, however, he has largely been forgotten by the layman. But to historians of science, Sir John Herschel is best understood, as Allan Chapman suggests, as "the living link between two styles or traditions of science" – the "independent gentleman" of science and the professional scientist.[51] Straddling as he did those two traditions, Herschel's Cape voyage casts important light on the varying roles of scientists in imperial contexts. The tension in Herschel's voyage helps illustrate Chapman's point: Herschel both rejected official help from the Admiralty because he could afford his own passage, while also taking full advantage of the political, military, and economic benefits the British Empire provided during his five years in the Cape colony to complete his observations and perform other scientific work of imperial benefit. Herschel's Cape voyage is, of course, only one of many episodes at the intersection of science and empire that challenge the idea that sciences, like astronomy, were unaffected by the circumstances in which they were practiced.[52] As this chapter has shown, Herschel's place at the pinnacle of Victorian science makes his Cape voyage, and publication of the *Cape Results*, uniquely informative, and hopefully encourages further research into the roles of scientists working in other imperial contexts.

Notes

1 William Herschel left £25,000 and property to John Herschel. See *Gentleman's Magazine*, 132 (1822): 650; and "Will of the late Sir William Herschel," *The Times* (October 9, 1822): 2. John Herschel to John William Lubbock, May 16, 1833, Herschel Papers, Royal Society, London, England (hereafter RS:HS): 21.136.

2 Quoted in Fergus Fleming, *Barrow's Boys: The Original Extreme Adventurers* (New York: Atlantic Monthly Press, 1998), 1.

3 "Review of Journal of an Expedition to explore the Course and Termination of the Niger; with a Narrative of a Voyage down that River to its Termination, by Richard and John Lander," *Edinburgh Review*, 55 (1832): 397.

4 Quoted in Sophie Waring, "The Board of Longitude and the Funding of Scientific Work: Negotiating Authority and Expertise in the Early Nineteenth Century," *Journal for Maritime Research*, 16 (2014): 55–71, on 63.

5 John Herschel to Francis Baily, April 24, 1832, RS:HS 21.106.

6 Basil Hall to John Herschel, September 3, 1832, RS:HS 9.171.

7 John Herschel to Basil Hall, September 16, 1832, RS:HS 21.115. Emphasis in original.

8 James Graham to Basil Hall, in a letter from Basil Hall to John Herschel, October 6, 1832, RS:HS 9.173.

9 John Herschel to John William Lubbock, May 16, 1833, RS:HS 21.136.

10 David S. Evans, Terence J. Deeming, Betty Hall Evans, and Stephen Goldfarb, eds., *Herschel at the Cape: Diaries and Correspondence of Sir John Herschel, 1834–1838* (Austin: University of Texas Press, 1969), 54.

11 Evans, *Herschel at the Cape*, 242 n. 46.

12 Evans, *Herschel at the Cape*, 47.

13 Evans, *Herschel at the Cape*, 48.

14 John Herschel to Margaret Herschel, July 23, 1830, quoted in Agnes Clerke, *The Herschels and Modern Astronomy* (New York: MacMillan and Company, 1895), 154.

15 Evans, *Herschel at the Cape*, 53.

16 Evans, *Herschel at the Cape*, 138.

17 Quoted in Brian Warner, "Sir John Herschel's Description of His 20-feet Reflector," *Vistas in Astronomy*, 23 (1979): 75–107, on 94.

18 See John Herschel, "Of the Distribution of Stars and of the Constitution of the Galaxy in the Southern Hemisphere," in *Results of Astronomical Observations Made During the Years 1834, 5, 6, 7, 8, at the Cape of Good Hope; Being the Completion of a Telescopic Survey of the Whole Surface of the Visible Heavens, Commenced in 1825* (London: Smith, Elder and Company, 1847): 373–92. Hereafter *Cape Results*. Compare to William

Herschel, "Astronomical Observations and Experiments Tending to Investigate the Local Arrangement of the Celestial Bodies in Space, and to Determine the Extent and Condition of the Milky Way," *Philosophical Transactions of the Royal Society of London*, 107 (1817): 302–31. On total stars tallied at the Cape, see Günther Buttmann, *The Shadow of the Telescope: A Biography of John Herschel* (Guildford: Lutterworth Press, 1974), 93.

19 John Herschel was the first to give the moons of Saturn their mythological names. See John Herschel, *Cape Results*, 415; Buttmann, *Shadow of the Telescope*, 106; and Omar W. Nasim, *Observing by Hand: Sketching the Nebulae in the Nineteenth Century*, (Chicago: University of Chicago Press, 2013). For the role of naming satellites in the context of contemporary controversies regarding Neptune, see Stephen Case, "A 'Confounded Scrape': John Herschel, Neptune, and Naming the Satellites of the Outer Solar System," *Journal for the History of Astronomy*, 50 (2019): 306–25.

20 Evans, *Herschel at the Cape*, 16 n. 37.

21 Steven Ruskin, *John Herschel's Cape Voyage: Private Science, Public Imagination and the Ambitions of Empire* (Aldershot, UK: Ashgate, 2004), 85

22 Spargo, P. E. "Foundations Strong and Lasting – Herschel's Work in Education at the Cape," *Transactions of the Royal Society of South Africa*, 49.1 (1994): 103–15, on 103–4. See also René Ferdinand Malan Immelman and William Thomson Ferguson, *Sir John Herschel and Education at the Cape, 1834 to 1840* (Oxford: Oxford University Press, 1961).

23 Spargo, "Foundations Strong and Lasting," 110, 112.

24 Brian Warner and John P. Burke, *Flora Herscheliana: Sir John and Lady Herschel at the Cape 1834 to 1838* (Johannesburg: Brenthurst Press, 1998).

25 Evans, *Herschel at the Cape*, 275; and Thomas Maclear, "Sir John Herschel at the Cape," *The Cape Monthly Magazine* (September 1871): 131–32.

26 Elizabeth Green Musselman, "Swords into Ploughshares: John Herschel's Progressive View of Astronomical and Imperial Governance," *British Journal for the History of Science*, 31 (1998): 419–36, on 431, 435.

27 J. L. Hilton, "The Herschel Obelisk, Classics, and Egyptomania at the Cape," *Akroterion*, 51 (2006): 117–33; Maclear, "Sir John Herschel at the Cape," 129.

28 Michael J. Crowe, *The Extraterrestrial Life Debate, 1750–1900: The Idea of a Plurality of Worlds from Kant to Lowell* (Cambridge: Cambridge University Press, 1988), 210.

29 Crowe, *Extraterrestrial Life Debate*, 213.

30 Thomas Paine, *The Age of Reason* (New York: Prometheus Books, 1984 [1793]), 59–60.

31 Crowe, *Extraterrestrial Life Debate*, 213–15.

32 Evans, *Herschel at the Cape*, 282.

33 Augustus de Morgan to John Herschel, December 30, 1842, RS.HS 6.188.

34 Buttmann, *The Shadow of the Telescope*, 119.

35 Herschel, *Cape Results*, 452.

36 Buttmann, *Shadow of the Telescope*, 121.

37 Michael Faraday to John Herschel, January 1, 1840, RS:HS 7.182.

38 Anonymous, "The Herschel Dinner," *Athenaeum*, June 16, 1838, 423.

39 John Herschel to James Forbes, March 30, 1839, RS:HS 22.5.

40 William Herschel, "Catalogue of Double Stars. By William Herschel, Esq. F. R. S." *Philosophical Transactions of the Royal Society of London*, 75 (1785): 40–126.

41 See Herschel, *Cape Results*, "Account of some Attempts to Compare the Intensities of Light of the Stars One with Another by the Intervention of the Moon, by the Aid of an Astrometer Adapted to that Purpose," 353–72.

42 As quoted in Michael J. Crowe, *Modern Theories of the Universe from Herschel to Hubble* (New York: Dover, 1994), 86.

43 Herschel, *Cape Results*, 389. For a thorough discussion of this topic, see Michael Hoskin, "John Herschel's Cosmology," *Journal for the History of Astronomy*, 18 (1987): 1–34.

44 David Brewster, "Review of *Cape Results*," *Littell's Living Age*, 16 (1848): 577.

45 Brian Warner, "The Years at the Cape of Good Hope," *John Herschel 1792–1871: A Bicentennial Commemoration* (London: The Royal Society, 1992): 51–66, on 62.

46 Herschel, *Cape Results*, 434.

47 For a thorough analysis of the reviews of the *Cape Results*, see Ruskin, *John Herschel's Cape Voyage*, 173–85.

48 Brewster, "Review," 577.

49 Walter F. Cannon, "John Herschel and the Idea of Science," *Journal of the History of Ideas*, 22 (1961): 215–39, on 217–18.

50 John Cawood, "The Magnetic Crusade: Science and Politics in Early Victorian Britain," *Isis*, 70.4 (1979): 492–518.

51 Allan Chapman, "An Occupation for an Independent Gentleman: Astronomy in the Life of John Herschel," *Vistas in Astronomy*, 36 (1993): 71–116, on 71.

52 See, for example, Robert A. Stafford, *Scientist of Empire: Sir Roderick Murchison, Scientific Exploration and Victorian Imperialism* (Cambridge: Cambridge University Press, 1989); David Philip Miller and Peter Hanns Reill, eds., *Visions of Empire: Voyages, Botany, and Representations of Nature* (New York: Cambridge University Press, 1996); Mark Harrison, "Science and the British Empire," *Isis*, 96 (2005): 56–63; and Andrew Goss, ed., *The Routledge Handbook of Science and Empire* (London: Routledge, 2021).

5

Herschel's Philosophy of Science

The Preliminary Discourse on the Study of Natural Philosophy and Beyond

Ever since its publication in 1830–31, John Herschel's *Preliminary Discourse on the Study of Natural Philosophy* has been recognized as epoch-making and canonical. It is rather surprising, therefore, that to date there exists little to no consensus on any aspect of the book, not even on what it is really about. This chapter provides a brief overview of the available literature on Herschel's *Preliminary Discourse*. In doing so, it attempts to both improve upon and go beyond current Herschel scholarship.

Herschel – the commentator on science – can be approached as a two-sided Baconian.[1] One the one side, like his close friend Charles Babbage (1791–1871), he was somewhat of a social reformer in the spirit of Bacon's *New Atlantis*, firmly convinced that science had the power to improve the human condition.[2] One the other side, like his other close friend William Whewell (1794–1866), Herschel was a methodologist in the spirit of Bacon's *Novum Organum*, seeking to present induction as the methodological motor of modern science. What made Herschel stand out was that he combined these two Baconian commitments to steer his own ('quietly utopian')[3] middle course between Babbage's technocratic rationalism and Whewell's conservative idealism. Herschel's Baconianism was oriented around the principle of the division of labour, applied to the various stages within the process of the production of scientific knowledge. Anyone could contribute to the elementary, observational stage of a discipline, but at the more advanced, theoretical stage its further development was reserved to major scientists. The *Preliminary Discourse* was focused almost entirely on the first stage, insisting that by engaging in this part of the process the public at large shared in the scientific mindset, which Herschel believed had a positive civilizing influence insofar as it promoted the virtues of self-restraint, modesty and piety.[4]

A fundamental tension ran through the methodological or philosophical side of Herschel's Baconianism, not unrelated to its sociopolitical side. It pertained to the advanced, theoretical stage of scientific development, where the Keplers and Newtons (and Herschels) of this world made their discoveries. The tension has often been described in technical terms as that between induction and hypothesis: in the *Preliminary Discourse*, Herschel articulated an inductivist philosophy of science – oriented around rule-bound observation open to all – while also seeming to allow for hypothetical speculation – acknowledging that major scientists should have free theoretical play. All commentators agree that Herschel rejected a rigid form of Baconian inductivism, while also recognizing that he stayed closer to tradition than Whewell. But whereas some argue that Herschel abandoned Baconianism entirely for a method of hypothesis, others insist that he held on to it by placing firm inductive restraints on hypothesizing. None, however, have taken much note of Herschel's philosophy of science beyond the *Preliminary Discourse*. Here, Herschel further developed a crucial aspect of his thinking only briefly touched upon in the *Preliminary Discourse*: namely, mathematics as a tool for scientific discovery and, more specifically, probability theory as a method for hypothesis-testing.[5] This not only provides new evidence for the interpretation of Herschel as a hypothetico-deductivist of sorts, but it also draws attention to the practice-oriented fashion in which Herschel transformed Baconianism to arrive at a fundamentally novel theory of scientific inquiry.

The chapter is divided into three sections. After introducing the origins of Herschel's *Preliminary Discourse*, it discusses the two major readings of the book, arguing that these are compatible and need to be combined. The discussion that follows orients around the need to reconcile Herschel's popular advocacy of inductivism with his practice-oriented hypothetico-deductivism. It puts forward the view that at the heart of the *Preliminary Discourse* stood a hierarchical division of scientific labour. On this basis, Herschel made it possible to argue that some low-level aspects of scientific inquiry are open to all, while reserving higher-level aspects to experts. The chapter then offers additional evidence for Herschel's endorsement of a qualified method of hypothesis for the advanced theoretical sciences, notably astronomy. This evidence suggests that Herschel came to look upon mathematics, and especially probability theory, as a tool for scientific discovery.

Preliminary Reflections on the *Preliminary Discourse*

Herschel's *Preliminary Discourse* was a steam-press produced, octavo book of some 350 pages in relatively cheap type selling for six shillings.[6] It was written over the course of the summer of 1830 and when published in 1830–31 was advertised in three formats: first, as part of Dionysius Lardner's *Cabinet Cyclopaedia*, one of many encyclopaedias that appeared in Britain after the pattern of the *Encylopédie*; second, as the lead work in its series on natural philosophy (Longman, the London publisher, sometimes announced it as 'A Preliminary Discourse for the Cabinet of Natural Philosophy'); and third, as an independent treatise. Every copy of the first edition included three title pages, allowing purchasers to decide for themselves whether they fitted the book into a series or library and, if so, which one. The first issue of seven thousand copies sold out in a few months; four thousand more were printed in 1831, and it went through several new but unrevised editions up to around 1851. Across the Atlantic, it was first published by Carey and Lea, Philadelphia, in 1831, and continued to be printed by Harper & Brothers, New York, up to 1869. During the second half of the nineteenth century, the book was translated into many languages, including French (1834), German (1836) and Italian (1840).[7]

Lardner's invitation to write the *Preliminary Discourse* came at a time when Herschel, aged thirty-seven, already had a high reputation among men of science but was not yet known as a writer for the general public. For Lardner, he was nonetheless the ideal author for the introductory work of his new encyclopaedia. Herschel had proven able to survey large and complex scientific subjects and had unusual expertise in all major scientific fields. Unlike some of his peers, notably Whewell – who hoped Herschel would 'give up spinning his entrails out into Encyclopaedias for such fellows as Lardner'[8] – Herschel was enthusiastic about widening his audience and, thereby, that of science itself. By 1830, he had already written lengthy contributions on 'Light' and 'Sound' for the *Encyclopaedia Metropolitana*, though these were introductory only to specialized students and never appeared separately. He had also agreed in 1829 to contribute a volume on astronomy to the *Cabinet Cyclopaedia*, which appeared in 1833 as *A Treatise on Astronomy*. For the *Preliminary Discourse*, his first popular book, Herschel took much care to present himself in a specific way: as a gentleman and as heir to Bacon and Newton. As James Secord has succinctly noted, Herschel evidently wanted his professional expertise to be put in the background, insisting that he appear on the title pages simply as 'John Frederick William Herschel, Esq., A.M., Late Fellow of St John's College, Cambridge, &c. &c. &c.', not trailed by 'F.R.S.' or any other honours and societies. For the frontispiece, Herschel chose an engraving of Francis Bacon's bust, and as the book's motto he used Aphorism 1 from Book I of the *Novum Organum*: 'Man, as the minister and interpreter of nature, is limited in act and understanding by his observation of the order of nature; neither his knowledge nor his power extends further.' Throughout the book, Herschel approvingly mentioned Bacon, 'our immortal countryman', twenty times and Newton some forty times. This way, Herschel kept alive the well-known seventeenth- and eighteenth-century image of Bacon as a prophetic Moses, seeing the promised land from afar, and Newton as Joshua, who took possession of it.

Much in the spirit of Bacon's 'Great Instauration', the *Preliminary Discourse* presented science as an ordered enterprise, in which the different parts of knowledge stand in clear relations to one another and to the whole. The book is divided into three parts. The first, 'On the General Nature and Advantages of the Study of the Physical Sciences' (74 pages), explains the purpose of scientific knowledge and the right frame of mind in which to pursue it. The second part (145 pages) examines the principles and rules guiding all physical sciences, richly illustrated with reference to their historical development. This is where Herschel discusses central issues like observation, classification, generalization, induction, hypothesis, analogy, laws, and causes, seeking to analyze and codify the scientific method. The final part, 'Of the Subdivision of Physics into Distinct Branches, and Their Mutual Relations' (140 pages), describes the major results achieved in the different branches of the physical sciences in terms of their objects of study, showing that scientific progress consists in 'continual transformation through discovery at the limits of knowledge'.[9]

Herschel's *Preliminary Discourse* 'represented a novel genre for the English reader'.[10] It not only surveyed the historical development of the sciences, but it also connected them to religious and social themes. For example, Herschel presented science as an essentially pious endeavour and emphasized the importance of its technological applications for the development of civilization, and more specifically the British Empire. Moreover, the account of the principles of natural knowledge was grounded not in abstract discussion of the laws of human thought but in a survey of the mental processes exhibited in scientific inquiry. As such, it formed an alternative both to the epistemological tradition, tracing from Locke and Hume to Reid, and to treatments of reasoning in general treatises on logic, such as those of Richard Whately. Herschel tellingly did not refer to any of these figures: in the book's fifty or so notes, references are to scientists (Hooke, Newton, Young, Fresnel, Lyell, etc.) rather than to philosophers, and one of the most oft-mentioned works, *The Four Ages* – which closely tied scientific progress to a broader process of civilization – was written by the eighteenth-century man-of-letters William Jackson (1730–1803).

Following Whewell, the *Preliminary Discourse* has often been lauded as 'the first book to root scientific progress systematically in its peculiar method'.[11] There is no doubt that Herschel's analysis of scientific methodology, which occupied the largest part of the book, put the topic firmly on the map in Britain, inspiring many others, scientists and philosophers alike, in several different ways. At the same time, the key to its wide appeal and success undoubtedly lay in the fact that it contained much more – indeed, almost 'everything which an intellectually cosmopolitan scientist thought to be relevant to an understanding of all the sciences'.[12] What then is the *Preliminary Discourse*? And how should it be read?

Reading the *Preliminary Discourse*

To date, there are two major readings of the *Preliminary Discourse*. The first and most dominant (R1) personifies the book as 'Herschel' and focuses on the *why* and *what* of the philosophical ideas on science it contains.[13] The second (R2) looks at its material form and orients around *how* it was written, published, read, and used – namely, decidedly *not* as a 'book of serious philosophy'.[14] Both readings are needed for a complete understanding of the *Preliminary Discourse*. Luckily, they are complementary and compatible, because they focus on different aspects of the book's origins and content – aspects that today may seem to have little or no connection but that were inextricably intertwined in the early Victorian period.

The Preliminary Discourse *as a Conduct Manual*

R2 draws attention to those parts of the *Preliminary Discourse* that Herschel scholarship, in its focus on the views on methodology put forward in the lengthy Part 2 (P2), has tended to neglect: Part 1 (P1) and, to a lesser extent, Part 3 (P3). This is where, in the words of Arthur Fine, Herschel brought his 'ode to the progress of science and humankind'.[15]

Chapter 1 ('Of man regarded as a creature of instinct, of reason, and speculation. – General influence of scientific pursuits on the mind') of P1 presents man as 'the undisputed lord of the creation' (3) for which the world is there to be used and known.[16] According to Herschel, scientific knowledge unavoidably leads to the recognition of 'order and design' (3) in nature and, through awareness of the fact that their origin is not human-made, to a conception of 'a Power and an Intelligence superior to his own' (3). As such, the common objection against the pursuit of science – namely, 'that it fosters in its cultivators ... self-conceit, leads them to doubt the immortality of the soul, and to scoff at revealed religion' (7) – is unfounded. Instead,

> Its natural effect ... on every *well constituted mind* is and must be the direct contrary. [It] must of necessity stop short of those truths which it is the object of revelation to make known; but, while it places the existence and principal attributes of a Deity on such grounds as to render doubt absurd and atheism ridiculous, it unquestionably opposes no natural or necessary obstacle to further progress: on the contrary, by cherishing as a vital principle an unbounded *spirit* of enquiry [science] unfetters the mind from prejudices of every kind, and leaves it open and free to every impression of a higher nature which it is susceptible of

> receiving, guarding only against enthusiasm and self-deception by a
> *habit* of strict investigation .... (7-8, **my emphasis**)

Just as science is not opposed to religion, insofar as 'truth can never be
opposed to truth' (9), it is also not reducible to its utilitarian use. On the
one hand, science is not a means to an end but a disinterested intellectual
pursuit valuable in itself. On the other hand, its history shows that
'speculations apparently the most unprofitable are almost invariably
those from which the greatest practical applications have emanated' (11).

Chapter 2 of P1 ('Of abstract science as a preparation for the study
of physics') introduces a division between abstract science (knowledge
of reasons and their conclusions) and natural or physical science
(knowledge of causes and their effects).[17] The former is based on
memory, thought and reason alone and concerned, first, with abstract
objects like space, time, number and order and, second, with artificial
symbols, which can be used as placeholders for these objects. For
instance, arithmetic and algebra are symbolic notations applied,
respectively, to number and order. In the *Preliminary Discourse*,
Herschel – contra the Cambridge tradition in which he himself was
trained and which he, along with other members of the Analytical
Society, helped innovate – deems acquaintance with such abstract
sciences profitable to but not necessary for the pursuit of physical
science. What suffices is the mere capacity to reason and recognize,
for example, that certain facts follow from a law of nature as its logical
consequences. This argument is crucial for Herschel's general message.
First, profound understanding of mathematics is 'altogether unattain-
able by the generality of mankind' (25). Second, 'no one need be
deterred from the acquisition of knowledge, or even from active original research in [physical] subjects, by a want of mathematical information' (26).[18] Herschel creates a tension here that runs through the
entire *Preliminary Discourse*. For in the same passage he admits that in
the case of the most advanced sciences – such as astronomy, optics and
dynamics, which are 'almost exclusively under the dominion of mathematics' (26) – original research is impossible without mathematical
proficiency. In other words, science is open to all, except when it is
reserved to some.[19] Herschel resolved this to some extent by introducing a (theoretically unstable and historically shifting) division of
labour between amateurs and experts.[20]

Chapter 3 of P1 discusses in more detail the topics of Chapter 1, oriented around the key notions of *law* and *cause*. It starts with an account of the nature and objects of science, both in itself and in its application to individual and social life. Herschel writes that rather than laying down particular laws, in creating the universe God impressed the 'permanent', 'consistent', 'intelligible' and 'discoverable' (42) laws that govern nature 'with the *spirit*, not the *letter*, of his law, and made all their subsequent combinations and relations inevitable consequences of this first impression' (37). The task for humans is 'to discover the *spirit* of the laws of nature ... from the *letter* which ... is presented to us by single phenomena' (39). This knowledge is important on many levels, from the improvement of 'the condition of mankind' (72) (e.g., vaccination) to the satisfaction of man's innate curiosity (e.g., discovery). Herschel repeatedly emphasizes that there exists a 'constant mutual exchange' between science and society, mediated by technology (i.e., '*scientific art*') (50) – a point almost entirely lost on someone like Whewell. Since Herschel defined science in terms of its special method, he was also able to suggest that the systematic and dispassionate frame of mind characteristic of scientists could be extended to 'the more complicated conduct of our social and moral relations' (73), such as in politics, legislation and economy. It was this point that the young John Stuart Mill (1806–73), in his review of the *Preliminary Discourse*, eagerly singled out as the book's central message.[21]

The bulk of P3 of the book – some 125 of a total of 140 pages – consists of a subdivision of the physical sciences into separate branches and a discussion of their mutual relations. Much like Bacon's writings on this subject,[22] Herschel's views on natural history have traditionally been largely neglected (both on R1 and R2). Natural history, as Herschel understood it, contains 'the elements of all our knowledge' and can be viewed in two ways: as 'the beginning or the end of physical science' (221). On the first view, it presents itself in the form of *classification*: the grouping of single phenomena into increasingly general classes (e.g., general facts, laws, axioms). This is one of the crucial steps in Herschel's account of scientific methodology, found in P2.[23] On the second view, pursued in P3, it takes the form of a subdivision of the sciences in terms of the results of their cause-effect analysis of 'natural agents' (e.g., force, heat). Over and above natural history stood

Herschel's unity-of-science thesis: all sciences deal with causes, operating as natural laws.[24] Since motion and force are the most general phenomena in nature, dynamics – the science of 'motion, as produced and modified by force' (223) – is placed at the head. Next to it stands statics, the study of forces acting on matter at rest under conditions of equilibrium. Each of these divisions of mechanics branch out into subdivisions, corresponding to the three different states of matter (solid, liquid or 'aëriform'), from the smallest (particles) to the largest (planetary systems).[25] Herschel's system had a number of striking features. For instance, geology was elevated to the level of astronomy, and mineralogy – one of the fields dealing with the 'material constituents of the world' (291) – was said to be the most connected science (geology, chemistry, optics, crystallography and physics).

The remaining fifteen pages of P3 contain reflections on the origins of modern science. Herschel asks why science emerged in Europe in the sixteenth century and not earlier or anywhere else in the world. The reasons he gives vary from the philosophical to the social and economic. There was a 'total want of a right direction given to enquiry, and of a clear perception of the objects to be aimed at, and the advantages to be gained by systematic and connected research' (348). Another reason was the 'increase in wealth and civilization' that made possible the 'necessary leisure' and helped spread the 'taste for intellectual pursuits', initially only in every 'principal European state' but eventually more and more in newly established 'civilized communities in every distant region' (349).[26] As the 'spirit of enquiry' (350) spread across the globe, there were more potential scientists who could contribute to a growing number of scientific fields (zoology, botany) and subjects (climates, tides, magnetic variations) in increasingly many ways (private collections, observations of local phenomena) and exchange and disseminate their findings more easily among each other (institutions, journals, proceedings). Other key factors of scientific development mentioned are the accumulation of accurate data and the improvement and construction of measurement instruments. These and other feats are deemed needed for the whole of humanity to pursue with success that 'infinite object' (360) called science.

On R2, looking through the lens of P1 and P2, the *Preliminary Discourse* is an affordable and accessible book on the human dimensions of science. Written at a time when the authority of science was attacked both from within (e.g., declinists) and without (e.g., members of the religious establishment), Herschel offered an apologia for and ode to its importance and significance for British society.[27] At the heart stood the idea, first, that science was open to all and, second, that its pursuit – either by *doing* it oneself or by *reading about* others doing it – nourished certain behaviour and mental habits (calm, disinterested, modest, truth-telling, pious). As such, it provided 'a foundation for good character across the social spectrum'.[28]

The Preliminary Discourse *as a Book on Philosophy of Science*

R1 focuses squarely on P2 of the *Preliminary Discourse*, where Herschel puts forward his philosophical views on scientific methodology. Rather than in the socio-political context of the 1830s, on R1 the book is placed in the tradition of eighteenth-century Scottish theoretical debates about science and ranked alongside Whewell's *History* (1837) and *Philosophy* (1840) and Mill's *System of Logic* (1843) as a classic of nineteenth-century English inductivist philosophy of science.[29]

During the 1960–70s, when the study of Victorian science took off, the *Preliminary Discourse* was seen as a statement of an early nineteenth-century consensus – that is, of the 'excessively Baconian cast of [early] Victorian notions of science'.[30] Somewhat later, the realization that 'Baconian' (and 'inductivist') was not a clear-cut, stable category gave rise to many different, even contradictory, interpretations of the *Preliminary Discourse*. Some argued that it was 'orthodox Baconian'[31] in its empiricist rejection of hypothesizing in science. Others suggested that it was straightforwardly 'anti-Baconian' or that in it Herschel had anticipated 'hypothetico-deductivism', allowing for non-rational guesses to hypotheses.[32] Historians are caught in a hall of mirrors when trying to answer whether or not Herschel's *Preliminary Discourse* was Baconian. First of all, the *Preliminary Discourse* was one of several publications from the early Victorian period that redefined the very meaning of this term; like Whewell, for instance, Herschel transformed the Baconian inheritance of which he claimed to be the legitimate heir.[33] Second, Herschel's philosophical outlook on science cannot, and should not, be identified entirely with the *Preliminary Discourse*.[34]

The disparate interpretations of Herschel's philosophy of science – a Baconian inductivist, an anti-Baconian hypothetico-deductivist or some midway position – are rooted in a tension running through the *Preliminary Discourse* itself.[35] This took the form of a division of scientific labour, which made the book's message appealing to two audiences. Since he sought to present science as accessible to all – while, as pointed out, it was in some cases actually reserved to some – Herschel had to carve out space for non-specialists[36] who lacked mathematical or scientific training. Herschel's famous 'rules of philosophizing' do precisely that, insisting that and stipulating how non-specialists can contribute to science: namely, by gathering, collecting and classifying data, essential for the specialist to discover laws and frame or test theories. At the same time, Herschel's liberality concerning hypothesizing – at least when compared to someone like Whewell – reflected a keen 'practitioner's awareness'[37] of the individual specialist's mature scientific practice. At some points, he seems to put inductive constraints on hypothesis-formation, suggesting that a hypothesis is legitimate only to the extent that it was framed on the basis of a body of already-acquired data. Elsewhere, however, he maintains that 'it matters little how [a hypothesis or theory] has been originally framed' (208) when it can withstand testing. The central question is whether Herschel can have his cake and eat it too – that is, whether he can abandon Bacon and remain a Baconian.

Herschel was influenced by Newton, but the *Preliminary Discourse* was indeed Baconian in character. P1 discussed such Baconian themes as the utility of knowledge and the extension of scientific methodology to the domains of politics and morality. P2 was framed as a renovation of Bacon's philosophy of science and more specifically his famous canons of induction. 'It is to our immortal countryman', Herschel wrote in Chapter 3 ('of the State of Physical Science in General, previous to the Age of Galileo and Bacon'),

> that we owe the broad announcement of this grand and fertile principle ... that the whole of natural philosophy consists entirely of a series of inductive generalizations...and a corresponding series of inverted reasoning from generals to particulars. (104)

According to Herschel, there had been no science, properly so called, before Bacon. Galileo, Copernicus and others showed that Aristotle was wrong, but it was Bacon who made clear *how* and *why* this was so. Moreover, Bacon replaced Aristotelian views on scientific inquiry with a 'stronger and better' methodology, oriented around cautious inductions, instead of 'deductions of principles of sweeping generality from few and ill-observed facts' (105). Bacon, therefore, 'will ... justly be looked upon in all future ages as the great reformer of philosophy' (115).

At the same time, though Herschel supported the spirit of the Baconian philosophical outlook, he was critical of the letter of Bacon's methodological writings. Herschel set out to revisit, once and for all, the major technical aspects of Baconian induction.

HERSCHEL'S TWO-SIDED PHILOSOPHY OF SCIENCE

Following Bacon, Herschel viewed induction as a stepwise and rule-bound inferential process involving 'the alternate use of both the *inductive* and *deductive* method' (174–75).[38] The goal of the inductive path is to arrive at explanations of observable phenomena in terms of proximate causes[39] or, if these cannot be established, laws – and, once these are available, at explanatory causal theories. More specifically, on the basis of the usual methods of observation, analysis, and classification (Chapters 4–5) and nine 'rules of philosophizing' (Chapter 6), scientists are (i) to derive and collect particular facts from observed phenomena; (ii) to analyze and classify these facts according to the analogies and resemblances among phenomena; (iii) to reduce complex phenomena into more elementary ones; (iv) to discover lower-level laws governing these general phenomena; (v) to obtain higher-level laws governing, or unifying, these lower-level laws.[40] Together, these steps make up the 'First Stage', focused on the isolation of a cause by removing or compensating for a confounding cause, mostly through experimental inquiry (Rule 7). At the 'Second Stage', scientists seek (vi) to formulate and verify explanatory theories of the causes of general phenomena (Chapter 7).

The deductive path of the inductive process is aimed primarily at verifying proposed causes or laws by testing them for empirical adequacy. 'If the empirical consequences of an inductively-grounded law or cause accommodate all known empirical data [and] predict novel phenomena...', as Aaron Cobb captures Herschel's reasoning, 'the evidence confirms the proposed law or cause'.[41] But according to Herschel, deductive methods also have other functions: the tracing of

empirical consequences can help scientists judge the accuracy of the
data used in the construction of laws, for instance, and disclose deeper
connections among phenomena, even if experience does not (dis)con-
firm the proposed law.

Importantly, as Cobb reminds us, the inductive and deductive paths
of 'Stage 1' and 'Stage 2' were said to be part of one and the same
inductive process. This explains, for instance, why Herschel could speak
of 'general laws *deduced* by legitimate *inductions* from observations.'[42]
It suggests that, far from abandoning Bacon, Herschel believed himself
to be re-defining Baconian induction: it contained everything from
observing and collecting and classifying data to following inductive
rules and using deductive methods.

Next to the inductive and deductive paths, Herschel recognizes the
value of hypothetical speculation in scientific inquiry, namely for the
search for ultimate causes of phenomena ('Stage 2').[43] Sometimes, he
argues that theoretical speculation always ought to be founded upon a
body of previously established inductive evidence ('Stage 1'). As such,
the formation of a theory or hypothesis ought not amount to 'the
unrestrained exercise of imagination, or [the] liberty to lay down
arbitrary principles, or assume the existence of mere fanciful causes'
(190). Or, that is, any hypothesis is legitimate if and only if the causes
it hypothesizes are actually *true causes* (*verae causae*).[44] At other
times, however, and apparently very non-Baconianesque, Herschel
downplays the cautious inductive ascent as a requirement for the
justification of higher-level theoretical (hypothetical, speculative or
conjectural) claims. Indeed, Herschel not only explicitly allows for
bold hypotheses as heuristic tools but also holds that in some cases
evidence can show that they (very likely) offer a true account of the
ultimate causes of phenomena, even when they posit unobservable
agents.[45] Any such hypothesis can legitimately be used by scientists in
the advanced, theoretical stage of their inquiry. Yet other passages
point to something of a middle way. Here, Herschel writes that
proposed hypotheses lacking any inductively established ground are
legitimate, albeit only when they, for instance, fit with existing empirical evidence and cohere with known laws.[46] Herschel uses a disturbing analogy: the freedom of hypothesizing is not like the 'wild license
of the slave broke loose from his fetters' but rather like that of the

'freeman who has learned the lessons of self-restraint in the school of just subordination' (190–91).

Taken together, Herschel might be seen as an inductivist who allows for hypothesizing only after, or at least congruous with, the acquisition of inductive evidence. But then again, there are statements in the *Preliminary Discourse* that support the view that Herschel adopted a qualified method of hypothesis.[47] John Dalton, for instance, is lauded for having arrived at his atomic theory on the basis of the consideration of only 'a few instances, without passing through subordinate stages of painful inductive ascent' (305–6). And there is also the more general claim that it sometimes 'matters little how [a theory] has been originally framed':

> However strange and, at first sight, inadmissible its postulates may appear . . . if they only lead us, by legitimate reasonings, to conclusions in exact accordance with numerous observations purposely made under such a variety of circumstances as fairly to embrace the whole range of the phenomena which the theory is intended to account for, we cannot refuse to admit them. (208–9)

In addition to the two general interpretations of Herschel's views on methodology – inductivist or hypothetico-deductivist – there is the interpretative puzzle of making sense of his endorsement of hypothetical inquiry. One option is to continue to focus on the inductivist sections of the *Preliminary Discourse* and dismiss the others as instances of creative indecision at best and ambiguity or sloppiness at worst. Another more fruitful and interesting option is to take the sections where Herschel qualifies his advocacy of inductivist constraints on hypothetical inquiry at face value and ask what these say about the *Preliminary Discourse* as a whole. By doing so, it becomes clear that R1 and R2 are not just compatible but complementary and that one should look beyond the *Preliminary Discourse* for a fuller picture of Herschel's philosophical outlook on science.

Re-reading the Preliminary Discourse

It was pointed out above that at various points in the *Preliminary Discourse* Herschel hinted at a division of scientific labour. On the one hand, there are novices without mathematical or scientific training.

(A broad category, containing everyone able to afford and read the book!) They can contribute to the immature observational sciences by following the inductive methods to obtain data needed by experts to (a) formulate theories and hypotheses that explain the data and (b) test these explanatory theories and hypotheses. On the other hand, there are experts. Only experts have the high-level mathematical expertise necessary to contribute to the ('highest') advanced theoretical sciences (e.g., astronomy and optics). Moreover, experts are free to infer bold hypotheses from an already-acquired body of inductive evidence provided that they extensively test their speculations.

R1 and R2 are both needed to understand this particular vision of science. R1 has traditionally focused on the philosophical ideas on inductive methodology found in the *Preliminary Discourse*, struggling to make sense of the anti-Baconian leanings of a self-proclaimed Baconian. By looking at the book's material form and narrative structure, R2 makes it possible to see that those passages that are clearly inductivist in nature – indeed, its very Baconianism – served a 'specific rhetorical purpose'.[48] The book was written for a non-specialist audience with the aim to promote the wholesome benefits of science for society and of scientific inquiry for practitioners. From a *reader's perspective*, therefore, Herschel had significant motivation 'to show that *some aspects* of scientific inquiry are open arenas for public participation and contribution'.[49] At the same time, Herschel's advocacy of hypothetical inquiry – of formulating and testing theories – represented a *practitioner's perspective* on science: experts should be alleviated as much as possible from the 'inferior' or 'mechanical' part of scientific labour to focus on the theoretical side of the subject at hand.[50] This gives a twist to R2's insistence that the *Preliminary Discourse* was not a 'recipe book'.[51] Herschel's point was not that following the rules of philosophizing (outlined in Part 2) would automatically lead to major discoveries. Neither did he want to suggest that great discoverers were to abide by a specifiable methodology, let alone low-level inductive rules cooked up by philosophers.

Within the intellectual landscape of the 1830s, discussions about scientific methodology were not yet separated from considerations about science's wider implications, from religion to morality. Herschel's *Preliminary Discourse* is a case in point. At its heart stands a division of scientific or mental labour that is hierarchical and at once scientific and social.[52] It is also unstable, since the common ground it creates ('everyone can do science') is premised on a division ('only some can do real science'). More precisely, what it says about the relationship between experts and method stands in tension with, or even threatens, the relationship between science and morality that the book seeks to establish.[53] Finally, it is also strategic. It opens to novices one part of the inferential process in some sciences. And it advocates a more hypothetico-deductive approach for breakthroughs in those sciences reserved to experts. This raises the question of what really constituted Herschel's view of what is distinctive about scientific inquiry, philosophically speaking.

Herschel's Philosophy of Science beyond the *Preliminary Discourse*

For an answer to this question one must look beyond the *Preliminary Discourse*. This is something that, perhaps surprisingly, very few scholars have done. Some of them have convincingly drawn from examples of Herschel's own experimental practice to discern a commitment to inductivist constraints for experts themselves. This shifts the argumentative burden to those who maintain that Herschel rejected inductivism for a qualified method of hypothesis. What follows offers pointers to additional evidence for the view that Herschel loosened the restrictions on hypothetical inquiry in mature or advanced scientific fields.[54]

The *Preliminary Discourse* was Herschel's first and only book on the philosophy of science. Over the course of the 1830s–50s he scribbled notes on inductive ethics and on a number of metaphysical themes (e.g., the nature of causality). Rather than from these archival sources, the evidence presented here derives mainly from published sources, all revolving around Herschel's advocacy of mathematics as a tool of scientific discovery, and more specifically of probability theory as a key example of the 'application of mathematics to natural philosophy' (219).[55] The focus will be on what this tells us about Herschel's philosophical outlook in and beyond the *Preliminary Discourse*. But it deserves mention that his methodological views developed in tandem with his astronomical studies, specifically relating to double stars and his activities as a 'business astronomer'.[56] This, in turn, sheds interesting new light on Herschel's apprenticeship with his father, William Herschel, which turns out to have been not only scientific but also philosophical in nature.

Probability in the Preliminary Discourse

Herschel introduced probability theory towards the end of Chapter 7 ('Of the higher degrees of inductive generalization, and of the formation and verification of theories') of P2. This is where he discusses 'Stage 2' of the inductive process – the stage of the 'exercise of pure reason' (190) at which explanatory theories and hypotheses are formulated and, more importantly, tested.

At the outset, Herschel claims that hypotheses should be founded upon a body of already acquired inductive evidence. He allows for bold hypotheses to function as heuristic aids in the search for explanatory theories of the ultimate causes of phenomena. In fact, he even holds that hypothesizing in combination with consequentialist testing makes for one way by which experts can discover 'general elementary laws' (199): form a bold hypothesis, particularize the law, and try the truth of it by following out its consequences and comparing them with facts (198–99). When a hypothesis stands the test of extensive verification, 'it matters little how it was originally framed' (208). Yet, Herschel admits that this is not the safest course for scientific inquiry. It is not always possible to achieve a genuinely crucial test to choose between theories, for instance (206–7). And when a theory is verified by comparing it with the facts, the appropriate kind of facts need to be available (208). Hence, exact physical data is key. For this, extensive and systematic observation is needed. But once established, the available data also needs to be 'well determined' (211). To have data is one thing; to know what one can learn from it is another. This is the point at which probability theory, and the theory of observational errors in particular, becomes important for Herschel.

When explanatory theories and hypotheses become more advanced, a more exact determination of data becomes necessary for their verification (see 213–14). 'But how,' Herschel asks, 'are we to ascertain . . . data more precise than observation itself?' (215) The answer: 'this useful and valuable property of the average of a great many observations, that it brings us nearer to the truth than any single observation can be relied on as doing' (215). Since probability theory is concerned with the calculation of 'the *probable* accuracy of our results, or the limits of error within which it is *probable* they lie', it stands at the basis of 'all physical enquiries where accuracy is desired' (215). It does more than make calculations upon observational data, however; it is also directly useful for physical theorizing:

> [Take] an observer – an astronomer for example – who would determine the exact place of a heavenly body. He points his telescope, and obtains a series of results disagreeing among themselves, but yet all agreeing within certain limits, and only a comparatively small number of them deviating considerably from the mean of all; and from these he is called upon to say, definitely, what he shall consider to have been the most probable place of his star at the moment. (218)

Probability is likely the most esoteric ('difficult and delicate') subject discussed in the *Preliminary Discourse*: a four-page discussion hidden in a chapter not really aimed at the book's primary readership. From the little he says about it, however, it is clear that Herschel attached great weight to it. This impression is confirmed by some of Herschel's later publications, briefly described below.

Probability beyond the Preliminary Discourse

Herschel opened his encyclopaedia article on 'Mathematics' from 1830 with a revealing statement about the nature of science, not found in the *Preliminary Discourse*: placing the advanced sciences under the label of '*mixed mathematics*', he writes that all such fields

> will participate in the uncertainty which must always hang about the inductive process, however far it may be pushed, since every conclusion which rests ultimately on the basis of observation must be infected with all the errors to which human observations, however carefully made, and however often repeated, are liable.[57]

Herschel continues by insisting that when such errors are due not to faulty reasoning but to 'a want of perfect accuracy in the observations', their effects can be 'correctly estimated'. Without explicitly mentioning it, Herschel undoubtedly had in mind probability theory when he wrote that this it is which renders the application of mathematical reasoning to natural science useful and satisfactory. . . It is always [possible] to assign limits beyond which the errors of observation cannot possibly extend [and] it is capable of demonstration that, in proportion as these limits are narrowed, the extent of the error induced on the result will experience a corresponding diminution.[58]

At the same time, the need of applying probability techniques to scientific inquiry also pointed to a more far-reaching conclusion: that all scientific knowledge is ultimately probable. 'It is always practicable', writes Herschel, 'to include such errors [of observation] in the expression of physical laws'; although 'the imperfection of our means of observation will not permit us to fix the precise point where truth resides, we are still enabled to say, with perfect certainty, that between such and such limits if must be found'.[59]

Herschel's only extensive discussion of probability appeared in print some twenty years later, in the form of a 100-page review (1850) of the works of Adolphe Quetelet (1796–1874) that can be taken as an 'addendum'[60] to the Preliminary Discourse. During the intervening two decades, when the study of probability and statistics in Britain expanded considerably, the topic occasionally featured in some of his correspondence and research. In 1845, for instance, Herschel wrote a long letter to John Stuart Mill in response to the latter's request for advice in preparing the second edition of his System of Logic.[61] Herschel objected to Mill's criticism of Pierre Simon Laplace's degree-of-belief interpretation of probability, especially the appeal to prior ignorance, and defended what would today be called a subjective Bayesian position: roughly, the idea that the probability of a hypothesis (e.g., about the existence of a cause) is the rational degree of belief that this hypothesis is true given the prior knowledge (and, ipso facto, ignorance) of the observer.

Already in 1832, Herschel read a paper in which he introduced a new method for finding the orbits of double stars, a line of research started by his father.[62] This graphic method, introduced to help determine 'the best approximation' of orbits from 'very poor data', had to do with probability insofar as its aim was to eliminate, or at least diminish, observational errors.[63] When properly applied, the astronomical observer would arrive at 'a conviction, almost approaching to a *moral certainty*, that we cannot be greatly misled'.[64]

A year later, in 1833, Herschel published *A Treatise on Astronomy* for Lardner's *Cabinet Cyclopaedia*, republished, with additions, as *Outlines of Astronomy* in 1849. Here, he devoted substantial space to double stars. Before it became possible to observe that double stars rotate around one another, the best evidence for a physical connection between them had been derived from a probabilistic argument: there were simply too many stars very close together for their distribution to be a matter of chance. This argument, first made by John Michell (1724–93) in 1767, had been referred to approvingly by Herschel's father. Herschel himself endorsed Michell's view in his 1849 book, arguing that it is possible to calculate the probability of two stars forming a binary pair when 'fortuitously scattered'. On the basis of new observational evidence, Herschel concluded that 'a physical connexion of some kind or other is ... unavoidable'.[65]

Herschel's argument sparked a controversy, in which some of Whewell's allies, notably James David Forbes (1809–68) and Robert Leslie Ellis (1817–59), took the lead.[66] All objected to the use of probabilistic arguments rather than inductions to infer the existence of causes. Ellis went as far as to claim that Herschel's attempt to turn 'the theory of probabilities [into] the philosophy of science, is in effect to destroy the philosophy of science altogether'.[67] Forbes' criticism was more precise. One of his arguments targeted Herschel's assumption that the 'doubt existing in the mind of a reasonable person' could be considered equivalent to an inherent probability. It is true that frequent close proximity of two stars can lead to an induction that might suggest a cause. But this could not be proved by probability alone. Forbes' other point was that Herschel had 'confused the measure of hypothetical antecedent probability of a given result with a probability in the nature of things ... and [had] used the measure of the former (which may be found correctly) for the measure of the latter'.[68]

Herschel responded to the criticism in his 1850 review of Quetelet's work, written for the non-specialist readers of the *Edinburgh Review*. Rather than going into detail, he chose to illustrate the use of probability in astronomy by means of an analogy he had already presented in the *Preliminary Discourse*. Shots fired at a wafer on a wall will distribute themselves in such a way that more shots hit near the wafer and fewer shots hit farther away. If the wafer is removed, how can it be determined from the marks alone where the wafer was located? This, Herschel argues, captures the problem that an astronomer faces: how to determine, from a set of many observations on a single star, the most probable location of the star? Herschel's answer was to use the method of least squares. Because the proof was too difficult for his readers, he had left it at that in the *Preliminary Discourse*. In 1850, he gave a simple derivation of the normal (or 'Gaussian') curve of errors to demonstrate how the shots should be distributed, which became known as 'Herschel's proof'.

Herschel further defended himself against Forbes (without mentioning him) by means of another analogy: take 'a target of vast size, marked out in 6700 millions of equi-distant rings'[69] and imagine placing in the bull's eye the number of stars within 4 seconds of arc of the nearest star, in the first ring stars within 8 seconds of arc of the nearest star, and so on. The result is a scattering that is abnormally distributed. Or, that is, the double stars near the bull's eye are connected by a cause, while the others are distributed entirely randomly.

Forbes had not, *pace* Herschel, denied the existence of double stars or claimed that probabilistic arguments are totally irrelevant. His point was that such arguments were never entirely conclusive. Herschel believed otherwise. Reflecting on his analogy, he wrote:

> Such we conceive to be the nature of the argument for a physical connexion between the individuals of a double star prior to the direct observation of their orbital motion round each. To us it appears conclusive .... We set out with a certain hypothesis as to the chances: granting which, we calculate the probability, not of one certain definite arrangement ... but of certain *ratios* being found to subsist between the cases in certain predicaments, on an average of great numbers.[70]

Like Bacon, both Herschel and Forbes believed that all natural knowledge is ultimately based on inductive inference. At the same time, they were worlds apart because each understood induction entirely differently. Forbes held that knowledge is conclusive only when it is acquired inductively through repeated observation. Unlike De Morgan or, somewhat later, William Stanley Jevons (1835–82), Herschel did not go so far as to suggest that induction is grounded on probability.[71] Instead, he took the view that no inductively verified, high-level knowledge is ever entirely conclusive: 'moral' or 'practical' certainty is the best that can be achieved.[72] And what else is this certainty than a conviction with a very high probability – a measure that can be quantified by means of observations? If it was not allowed to call this position Baconian, then so much the worse for Baconianism.

Notes

1 Here, I follow Rose-Mary Sargent, in Francis Bacon, *Selected Philosophical Works*. Edited, with Introduction, by Rose-Mary Sargent (Cambridge: Hackett, 1999), xxviii.

2 This side of Herschel's Baconianism is set out in William Ashworth's contribution to this volume (Chapter 10). The exact connection between the two parts is an open question and, arguably, an interesting one for future research.

3 James A. Secord, *Visions of Science: Books and Readers at the Dawn of the Victorian Age* (Chicago: University of Chicago Press, 2014), 89.

4 For a stimulating recent account of the 'epistemic virtues' of early nineteenth-century British science, see Richard Bellon, 'Sacrifice in Service to Truth: The Epistemic Virtues of Victorian Britain', in Emanuele Ratti and Thomas A. Stapleford (eds.), *Science, Technology, & Virtues: Contemporary Perspectives* (Oxford: Oxford University Press, 2021), 17–27.

5 A perceptive but little cited essay in this regard is Marie Boas Hall, 'The Inductive Sciences in Nineteenth-Century England', in Trevor H. Levere and William R. Shea (eds.), *Nature, Experiment, and the Sciences* (Dordrecht: Kluwer, 1990), 225–47, see 225.

6 On the material dimension of Herschel's *Preliminary Discourse* see Secord, *Visions of Science*, ch. 3; and Michael Partridge, 'Introduction', in John F. W. Herschel (ed.), *A Preliminary Discourse on the Study of Natural Philosophy* (London: Johnson Reprint, 1966). Herschel was invited to write the introduction to the 'cabinets' on natural philosophy on 19 January 1830 by the editor, Dionysius Lardner, who offered him £250. Herschel accepted, writing on 2 February that he expected to finish the book in about six months. (See RS:HS 11.115 and RS:HS 11.116.) Lardner shared the first proofs of the *Preliminary Discourse* with Herschel early in October 1830 (RS:HS 11.119). Herschel's diary shows that he submitted his last revisions on 17 December 1830, which means that the first print rolled off the press within two weeks' time. See Diary, 17 December 1830, Herschel Family Papers, Harry Ransom Center, University of Texas, W0012, container 16.17.

7 For details on the various translations of Herschel's *Preliminary Discourse* see Michael J. Crowe, 'Bibliography of the Publications of Sir John Herschel', in Brian Warner (ed.), *John Herschel, 1792–1992: Bicentennial*

Symposium, vol. 49, part I (Cape Town: Royal Society of South Africa, 1994), 125–40, on 129.

8 William Whewell to Richard Jones, [November 1830], Trinity College Library, Cambridge, Whewell Papers, Add.Ms.c.51/91.

9 Secord, *Visions of Science*, 90.

10 Richard Yeo, 'Reviewing Herschel's *Discourse*', *Studies in History and Philosophy of Science Part A*, 20.4 (1989): 541–52, on 545.

11 Whewell (1831) paraphrased in Henry Cowles, 'The Age of Methods: William Whewell, Charles Peirce, and Scientific Kinds', *Isis*, 107.4 (2016): 722–37, on 726.

12 Walton F. Cannon, 'John Herschel and the Idea of Science', *Journal of the History of Ideas*, 22.2 (1961): 215–39, on 222.

13 This reading, which traces back to Whewell and Mill, emerged in the 1960s and has since that time given rise to a significant body of literature, prominent instances of which are (in chronological order): C. J. Ducasse, 'John F. W. Herschel's Methods of Experimental; Inquiry', in R. M. Blake, C. J. Ducasse, and Edward H. Madden (eds.), *Theories of Scientific Method: The Renaissance through the Nineteenth Century* (Seattle: University of Washington Press, 1960), 153–82; Cannon, 'John Herschel and the Idea of Science', David B. Wilson, 'Herschel's and Whewell's Versions of Newtonianism', *Journal of the History of Science*, 35 (1974): 79–97; Chaman Lal Jain, 'Methodology and Epistemology: An Examination of Sir John Frederick William Herschel's Philosophy of Science with Reference to His Theory of Knowledge', PhD diss., Indiana University, 1975; John V. Strong, 'John Stuart Mill, John Herschel, and the "Probability of Causes"', *PSA: Proceedings of the Biennial Meeting of the Philosophy of Science Association*, 1 (1978): 31–41; Gregory Good, 'John Herschel's Optical Researches and the Development of His Ideas on Method and Causality', *Studies in History and Philosophy of Science* 18 (1987): 1–41; Marvin Paul Bolt, 'John Herschel's Natural Philosophy: On the Knowing of Nature and the Nature of Knowing in Early Nineteenth-Century Britain', PhD diss., University of Notre Dame, 1998; Laura J. Snyder, 'Hypotheses in 19th-century British Philosophy of Science: Herschel, Whewell, Mill', in Michael Heidelberger and Gregor Schiemann (eds.), *The Significance of the Hypothetical in the Natural Sciences* (New York: Walter de Gruyter, 2009), 59–76; Aaron D. Cobb, 'Is John F. W. Herschel an Inductivist about Hypothetical Inquiry?', *Perspectives on Science*, 20.4 (2012): 409–39.

14 Secord, *Visions of Science*, 82.

15 Arthur Fine, 'Foreword', in John F. W. Herschel (ed.), *Preliminary Discourse on the Study of Natural Philosophy*. Facsimile of the first edition, 1830, with a new Foreword by Arthur Fine (Chicago: University of Chicago Press, 1987), xi.

16 In what follows, page numbers between brackets refer to Herschel's *Preliminary Discourse*, unless otherwise noted.

17 Throughout the *Preliminary Discourse*, Herschel used 'sciences' in the sense of the physical sciences, excluding the abstract sciences as well as the sciences dealing with animate subjects (cf. n. 25).

18 In anticipation of the considerations put forward in "Re-reading the *Preliminary Discourse*," compare this statement with Herschel's claim that 'to a very moderate progress in [advanced theoretical sciences or mixed mathematics], a very perfect knowledge of [pure mathematics] is requisite'. John F. W. Herschel, 'Mathematics', in David Brewster (ed.), *Edinburgh Encyclopedia* (Edinburgh: W. Blackwood, 1830), vol. 13, 1, 359–83, on 360.

19 Herschel's work on astronomy arguably offers an example of this hierarchical dichotomy. For instance, he sent his fellow observers directions on how to make double-star observations, but he used those himself to calculate their orbits. See Stephen Case, *Making Stars Physical: The Astronomy of Sir John Herschel* (Pittsburgh: University of Pittsburgh Press, 2018), 75–83.

20 Cf. Yeo, 'Reviewing Herschel's *Discourse*', 551; Timothy L. Alborn, 'The Business of Induction: Industry and Genius in the Language of British Scientific Reform, 1820–1840', *History of Science*, 34.1 (1996): 91–121.

21 See John Stuart Mill, 'Herschel's *Discourse*', *Examiner*, 20 March 1831, 179–80.

22 See Dana Jalobeanu, *The Art of Experimental Natural History: Francis Bacon in Context* (Bucharest: Zeta Books, 2015), Introduction.

23 Herschel, *Preliminary Discourse*, Part 2, Chapter 5 ('Of the Classification of Natural Objects and Phenomena, and of Nomenclature'). See also Mary Poovey, *A History of the Modern Fact: Problems of Knowledge in the Sciences of Wealth and Society* (Chicago: University of Chicago Press, 1998), 317–19.

24 This separates Herschel's *Preliminary Discourse* from another work appearing in Lardner's *Cabinet Cyclopaedia*: Baden Powell, *An Historical View of the Progress of the Physical and Mathematical Sciences, from the Earliest Ages to the Present Times* (London: Longman, Rees, Orme, Brown, Green, & Longman, 1834).

25 Herschel had very little to say about sciences dealing with animate subjects, such as biology, zoology and botany: 'They form, it is true, a most important and deeply interesting province of philosophical enquiry; but the view that we have taken of physical science has rather been directed to the study of inanimate nature, than to that of mysterious phenomena of organization and life' (343).

26 On aspects of the intimate connection between science and the British Empire in the early nineteenth century see, for instance, John Gascoigne, 'Science and the British Empire from Its Beginnings to 1850', in Brett M.

Bennett and Joseph M. Hodge (eds.), *Science and Empire: Knowledge and Networks of Science across the British Empire, 1800–1970* (Basingstoke: Palgrave Macmillan, 2011), 47–67.

27 On aspects of the broader context in which the *Preliminary Discourse* appeared, see, for instance, Secord, *Visions of Science*, 87–91; and Yeo, *Defining Science*, ch. 2. One of Secord's arguments is that Herschel wrote the *Preliminary Discourse* in response to his friend Charles Babbage's *Reflections on the Decline of Science in England* (1830) – with whose reforming aims he agreed but whose polemical tone and political edge he disliked. See Secord, *Visions of Science*, 87–89.

28 Secord, *Visions of Science*, 82.

29 The sources of Herschel's philosophical views on science remain an open question. Olson was the first (and, to date, the only one) to approach this topic by placing the *Preliminary Discourse* in the context of the Scottish Common Sense tradition. While striking, and well worth further study, the evidence for the connection is rather indirect. See Richard S. Olson, *Scottish Philosophy and British Physics, 1740–1870: A Study in the Foundations of the Victorian Scientific Style* (Princeton: Princeton University Press, 1975), ch. 10.

30 Charles Coulston Gillispie, *The Edge of Objectivity: An Essay in the History of Scientific Ideas*, with a new introduction by Theodore M. Porter (Princeton: Princeton University Press, 2016 [1960]), 314.

31 See, for instance, John Agassi, 'Sir John Herschel's Philosophy of Science', *Historical Studies in the Physical Sciences*, 1 (1971): 1–37, especially 1–2, 20–21. Agassi is probably the only scholar who has interpreted Herschel as an advocate of a strict form of inductivism.

32 See, respectively, Walter P. Cannon, 'History in Depth', *History of Science*, 3 (1964): 20–38; and Larry Laudan, *Science and Hypothesis: Historical Essays on Scientific Methodology* (Dordrecht: D. Reidel, 1981), 129–31, 189. For this interpretation, see also, for instance Geoffrey Cantor, 'The Reception of the Wave Theory of Light: A Case Study Illuminating the Role of Methodology in Scientific Debate', *Historical Studies in the Physical Sciences*, 6 (1975): 109–32; and Daniel M. Siegel, *Innovation in Maxwell's Electromagnetic Theory: Molecular Vortices, Displacement Current, and Light* (Cambridge: Cambridge University Press, 1991), ch. 1.

33 For a discussion of the Herschel-Whewell-Mill connection – relating to their struggle over who was Bacon's legitimate heir – see Laura J. Snyder, *Reforming Philosophy: A Victorian Debate on Science and Society* (Chicago: University of Chicago Press, 2006), chs. 1 and 2; and Snyder, 'Hypotheses in 19th Century British Philosophy of Science'. On the history of Baconianism and Bacon's reception in the nineteenth century, see Yeo, 'An Idol of the Market-Place'; Lukas M. Verburgt, '*The Works of Francis Bacon*: A Victorian Classic in the History of Science', *Isis*, 112.4 (2021): 717–36; and

Lukas M. Verburgt, 'Scientific Method, Induction and Probability: The Whewell-De Morgan Debate on Baconianism, 1830–1850', forthcoming in *HOPOS: The Journal of the International Society for the History of Philosophy of Science.*

34 This point will be explored in the next section.

35 To identify this tension is, of course, to take an opinionated view on the *Preliminary Discourse.* What follows is largely, though not entirely, in line with the interpretation put forward in Bolt, 'John Herschel's Natural Philosophy'. It diverges from Bolt, and agrees, for instance, with Snyder, in insisting that Herschel redefined rather than abandoned Baconianism.

36 The oft-used 'amateurs' would not be the right expression, as Herschel liked to think of himself as an amateur scientist.

37 Cobb, 'Is John F. W. Herschel an Inductivist about Hypothetical Inquiry?', 410.

38 The account of Herschel's philosophy of science offered here (pp. 110–112) largely follows Cobb's presentation, especially his distinction between (different aspects of) 'Stage 1' and 'Stage 2'. See Cobb, 'Is John F.W. Herschel an Inductivist about Hypothetical Inquiry?'.

39 Throughout the *Preliminary Discourse*, Herschel uses the term 'cause' in at least four different senses. See Ducasse, 'Herschel's Methods of Experimental Inquiry'. In what follows, 'cause' is used in the sense of an antecedent phenomenon of an observed regularity, unless otherwise stated. However, while acknowledging the cognitive limitations of human knowledge, Herschel did believe that science ought to seek the ultimate causes of phenomena.

40 See Cobb, 'Is John F.W. Herschel an Inductivist about Hypothetical Inquiry?', 414.

41 Cobb, 'Is John F.W. Herschel an Inductivist about Hypothetical Inquiry?', 417.

42 Herschel, 'Mathematics', 359.

43 Herschel recognizes that the human propensity to speculate is also apparent in 'Stage 1' of the inductive process, where 'on the least idea of an analogy between a few phenomena, it leaps forward . . . to a cause or law' (164). This is, in part, why he advocates the alternating use of inductive and deductive methods at this stage, aimed at disciplining the tendency, so to speak. In this sense, probability tests can be seen as the kind of disciplining appropriate to 'Stage 2'.

44 The *vera causa* principle is a much discussed aspect of Herschel's inductivism. In short, it says that a cause is real if it is sufficient to account for the phenomenon in question.

45 See Cobb, 'Is John F.W. Herschel an Inductivist about Hypothetical Inquiry?', 420–1.

46 See ibid., 422.

47 The most substantive defense of this interpretation is provided in Bolt, 'John Herschel's Natural Philosophy'.

48 Cobb, 'Is John F. W. Herschel an Inductivist about Hypothetical Inquiry?', 423.

49 Cobb, 'Is John F. W. Herschel an Inductivist about Hypothetical Inquiry?', 423.

50 For 'inferior' see Herschel, *Preliminary Discourse*, 191; for 'mechanical' see John F. W. Herschel, 'Address', 19 June 1845, *Report of the Fifteenth Meeting of the British Association for the Advancement of Science; Held at Cambridge in June 1845* (London: John Murray, 1846), xxvii–xliv, on xxxiv.

51 Secord, *Visions of Science*, 91.

52 Cf. Ashworth, 'The Calculating Eye'; and Alborn, 'The Business of Induction'.

53 On the complex and shifting connections between these categories in the early Victorian period, see Richard Yeo, 'Genius, Method, and Morality: Images of Newton in Britain, 1760–1860', *Science in Context*, 2.2 (1988): 257–84. Unfortunately, Herschel is not considered in Yeo's otherwise brilliant analysis.

54 Cf. Aaron D. Cobb, 'Inductivism in Practice: Experiment in John Herschel's Philosophy of Science', *HOPOS: The Journal of the International Society for the History of Philosophy of Science*, 2.1 (2012): 21–54; Steven Ruskin, *John Herschel's Cape Voyage: Private Science, Public Imagination and the Ambitions of Empire* (Aldershot, UK: Ashgate, 2004), 'Part II: The Production of the *Cape Results*'.

55 Herschel, indeed, was the first among British men of science to approach probability this way, rather than as a branch of mathematics, as, for instance, Augustus De Morgan did in the 1830s.

56 Ashworth introduced this phrase to refer to the loosely knit group of mathematicians – to which belonged, among others, Herschel, Charles Babbage, John Lubbock and Augustus De Morgan – involved in the founding of the London Astronomical Society. See William J. Ashworth, 'The Calculating Eye: Baily, Herschel, Babbage and the Business of Astronomy', *The British Journal for the History of Science*, 27.4 (1994): 409–41.

57 Herschel, 'Mathematics', 359.

58 Herschel, 'Mathematics', 360.

59 Herschel, 'Mathematics', 360.

60 For this observation, see Thomas L. Hankins, 'A "Large and Graceful Sinuosity": John Herschel's Graphical Method', *Isis*, 97.4 (2006): 605–33, on 606.

61 On the Mill-Herschel exchange on probability, see Strong, 'Probability of Causes'; and Theodore M. Porter, *The Rise of Statistical Thinking, 1820–1900* (Princeton: Princeton University, 2020 [1986]), 82–83.

Herschel's letters to Mill can be found in *The Earlier Letters of John Stuart Mill, 1812–1848*, 2 vols., ed. F. E. Mineka (Toronto: University of Toronto Press, 1963).

62 For details, see Hankins, 'John Herschel's Graphical Method'; and Case, *Making Stars Physical*, ch. 3.

63 Hankins, 'John Herschel's Graphical Method', 609.

64 John F. W. Herschel, 'On the Investigation of the Orbits of Revolving Double Stars. Read 13 Jan. 1832', *Memoirs of the Royal Astronomical Society*, 5 (1833): 171–222, on 179. The brief discussion of Herschel's views on astronomy and probability offered here (pp. 118–19) follows Hankins, 'John Herschel's Graphical Method', especially pp. 625–30.

65 John F. W. Herschel, *Outlines of Astronomy*, 4th ed. (London: Longman, Brown, Green, and Longmans, 1851 [1849]), 565. On Herschel's reasoning, in this regard, and the controversy it elicited see, for instance, Barry Gower, 'Astronomy and Probability: Forbes versus Michell on the Distribution of the Stars', *Annals of Science*, 39 (1982): 145–60; and Lukas M. Verburgt, 'A Letter of Robert Leslie Ellis to William Walton on Probability', *Journal of the British Society for the History of Mathematics*, 33.2 (2018): 96–108.

66 For a more detailed discussion of this controversy, see Hankins, 'John Herschel's Graphical Method', 628–30.

67 Robert Leslie Ellis to James David Forbes, 3 September 1850. Ellis belonged to the small circle of colleagues with whom Forbes shared a draft of his criticism of Herschel's argument about double stars, published as James David Forbes, 'On the Alleged Evidence for a Physical Connexion between Stars Forming Binary or Multiple Groups, Arising from Their Proximity Alone', *Philosophical Magazine*, 35 (1849): 132–33.

68 Forbes, 'On the Alleged Evidence for a Physical Connexion', 416.

69 John F. W. Herschel, 'Quetelet on Probabilities', *Edinburgh Review* 92 (1850): 1–57, on 36. Herschel made a mistake in deriving the law for the normal distribution of shots at a target, which he acknowledged in a popular paper from 1867, 'On the Estimation of Skill in Target Shooting'. See, in this regard, Hankins, 'John Herschel's Graphical Method', 630.

70 Herschel, 'Quetelet on Probabilities', 37.

71 See, in this regard, Larry Laudan, 'A Note on Induction and Probability in the 19th Century', in *Science and Hypothesis: Historical Essays on Scientific Methodology* (Dordrecht; Boston: Reidel, 1981), 192–201.

72 These phrases appear, respectively, in Herschel, 'On the Investigation of the Orbits of Revolving Double Stars', 179, 181; Herschel, 'Quetelet on Probabilities', 2–3, 20.

OMAR W. NASIM

6

Drawing Observations Together

John Herschel and the Art of Drawing in Scientific Observations

As early as 1826, John Herschel wrote, "astronomers are seldom draftsmen, and have hitherto ... contented themselves with very general and hasty sketches."[1] This was more than a passing complaint with past depictions of astronomical objects. It was an exhortation to a generation of astronomers to begin taking draftsmanship seriously. With its increasing focus on the physical features of astronomical bodies like the Sun, the planets, comets, and the Moon, as well as the intricate physical complexities of the Milky Way, celestial nebulae, and clusters of stars, the astronomy of the nineteenth century demanded a new and systematic focus on draftsmanship as a means of observation. In this way, astronomy was very much in line with other observational sciences of the period like geology and minerology, botany and zoology, archaeology and ethnography, all of which required, alongside detailed descriptions, more exact pictures of their respective subject matter.[2] As the act of drawing became embedded into routine scientific record-keeping practices, it also became closely tied to what counted as proper scientific observation. Herschel was not just in tune with these important developments, he exemplified them in his own observational performances. This is evinced by his exquisite skills and techniques in drawing in the service of science and, above all, astronomy. Herschel's observational practices were embodied in a set of visualizing instruments, techniques, and materials. What follows is a survey of some of these practices as they range over a lifetime of carefully observing and drawing many sorts of phenomena.

Visual studies, broadly construed, has taught us to regard all sorts of visual representations as entry points to the many ways the world has been perceived or experienced. Since there are multiple ways the world has been visually represented in history, each with its own rich contexts and circumstances, these, it stands to reason, afford us diverse histories of perception.[3] By looking at how Herschel produced detailed drawings of the world, we may be able to say something about what counted as perception for him and his time. But I will go a step further in this survey of Herschel's drawing practices. I take the means of their production, including varied instrumental aides seriously as clues to, if not indicators of broader contexts: if a visual record is an indication of not just *what* someone saw but *how* they looked at something, then the instruments used in the production of that record are also instruments of perception. And if we take observation as a more refined form of perception, as we shall see Herschel does, then we must also be attuned to the ways such refinement was attained for the purposes of a science. These means will be intellectual but also material, practical, and technical (in the sense of technique). They are inscribed into the scientific record and the very form such records took in order to be readable to the scientific gaze.

In what follows, therefore, we take time to situate Herschel's drawing practices and techniques into the context of the early nineteenth century and its developments concerning drawing implements and their connections to scientific observation. These include the modern graphite pencil, wove paper, "working skeletons," the camera lucida. From the emergence of these drawing instruments and the novel attitudes and aims they underwrote, we follow some of the ways Herschel's own practices compared with his contemporaries. With this backdrop of contextualized similarities and differences, we identify a mode of observation typical to Herschel, one that extends from his early drawings of archaeological ruins and geological features all the way to his extensive drawings of challenging celestial objects like nebulae. This highly refined form of observation disciplined the act of drawing to be one of its primary instruments of perception. In this chapter, I situate and pinpoint what was common to many of his observational practices, especially those that employed drawings. By doing so, I draw together nearly a lifetime of Herschel's observational performances.

Outdoors with Paper and Pencil

John Herschel was likely tutored to draw from an early age, as was the custom for the children of well-to-do families at the turn of the century.[4] Indeed, it has been claimed that Herschel "had become an accomplished draftsman while still a child."[5] In addition to the life-long stipend guaranteed by King George III to John Herschel's father and the latter's Europe-wide telescope business, Sir William Herschel's marriage to Mary Pitt (former Baldwin), the widow of the wealthy London merchant John Pitt, brought financial stability to generations of the Herschel household. These circumstances secured the best education for John at the time: first at Dr. Gretton's private school at Hitcham, Buckhinghamshire, then for a few months at Eton, and afterwards at home by private tutors (see Chapter 1).[6] Normally a part of a privileged education, John Herschel was likely taught drawing in all these contexts. By the time Herschel entered Cambridge University, his aunt Caroline Herschel could proudly declare that "I heard Hauptman Müller wishing to have but one of John's Talents, viz. that of Drawing."[7] What may be Herschel's earliest surviving drawing (pen, pencil and wash) is found in a page of a diary from August 4, 1809, when he was seventeen years old.[8] Just a year later, on a late July trip to the Lake District, Herschel produced at least twenty-two individual drawings of different lakes in the region, including the Windermere, Derwent Water, and Ullswater. These were no longer relatively quick entries into a diary but stand-alone works, done on individual sheets of wove paper.

We know of these early drawings thanks to a valuable document Herschel created near the end of his life entitled *General List of all my Drawings and Sketches Made in July 1861* (though it includes drawings made by him as late as 1870, a year before his death).[9] The list comprises a total of 763 drawings, mostly in pencil, made over the course of Herschel's lifetime, including 260 drawings from at least four Grand Tours (1821, 1822, 1824, 1826); 132 landscapes done at the Cape of Good Hope (1834–38) where he, with his wife, also made 132 botanical drawings; and about 98 sketches made on tours through Great Britain. Both the Grand Tour excursions and the Cape drawings include a number of those dedicated to geological and topographical phenomena. Also contained in this list is around 82 drawings of astronomical objects, including the Milky Way, Halley's Comet, M42 (the nebula in Orion), and Eta Argus (today Eta Carinae). As large as the inventory is, however, it does *not* include hundreds of sketches in different states of completion found scattered throughout Herschel's many notebooks and diaries or the many loose sheet drawings that exist in folders and albums throughout his extensive and dispersed archive. There can be no question that Herschel was a prolific draftsman, with expert abilities in drawing and sketching all kinds of subject matter, natural and cultural. That he fluently employed the pencil to depict such a wide range of scientific and aesthetic objects speaks to the substantial value he placed on the art of drawing.[10]

The majority of drawings recorded in the *General List* were made with instruments and attitudes that were very much of their time. For instance, Herschel seems to have readily adopted the then-recent tendency among artists to draw *en plein air* rather than just in their studios.[11] This was an idea influentially articulated by the French artist Pierre-Henri de Valenciennes (1750–1819) in 1800 and fervently taken up by artists like Achille Etna Michallon (1796–1822), Jean-Baptiste Camille Corot (1796–1875), Simon Denis (1755–1813), John Constable (1776–1837), William Turner of Oxford (1789–1862), George Bryant Campion (1795–1870), and others.[12] Whether on the edge of a cliff, in an open valley, or in the thicket of a forest, this approach to drawing required mobility and portability of seat and easel or drawing table. And unlike in the studio, where artists were notoriously prone to stylize, the outdoors forced a kind of naturalistic precision on artists: "I also see how exacting," Corot declared to his teacher Michallon, "we must be in drawing from life and above all not to be satisfied with a hastily executed sketch."[13] Generalization and haste were now the adversary. And when confronted with the unpredictability of Northern European weather, outdoor observers required durable drawing instruments and tools. Like Corot, for instance, Herschel much preferred to use the versatile, durable, and portable pencil, rather than, say, watercolors, oils, or wash.

The modern pencil itself was an important technological advancement. Formulated in direct competition to the ancient and well-established plumbago, an earlier and much older form of the pencil made with English Cumberland graphite, the French artist and polymath Nicolas-Jacques Conté (1755–1805) fabricated the pencil in 1795 by mixing finely powdered graphite and potter's clay and fusing this mixture in high-temperature furnaces. The mixture was then poured still hot into molds, producing thin ceramic sticks of lead when cooled. These sticks were encased in wooden surrounds, affording them durability and making it possible to sharpen the lead into sharp points. Unlike the plumbago, what made Conté's pencils particularly unique and useful was the wide range of tonal and textural variation that could be attained by standardized modifications to the production process. Pencils could be (and have been since the time of Conté) divided and sold in two groups: "technical" and "artistic," with four kinds of pencils manufactured in each group with their own distinct qualities. The division between these two general categories, explains one art historian, was "not just of use, but of effect, for the first set ['technical'] … suggested a gradation from softness to hardness, whereas the second, 'artistic' set tracked a tonal spectrum characterized by the relative darkness of the marks made."[14] It is clear, even on a cursory examination of Herschel's drawings, that he often alternated between artistic and technical pencils, using one set for shadows and texture and another set for its bold and hard lines. Herschel, that is, made full use of the range afforded by the modern pencil for the rich detail and nuanced tonal variations so characteristic of his drawings (see Figure 6.1).

The wood-encased pencil, with its differentiated assortment, portability, and durability, encouraged a new tendency in the visualization of the world at the start of the nineteenth century, a tendency also found in Herschel's own visual procedures. The art historian Joseph Meder explains that with Conté pencils, which would subsequently be manufactured by firms like the famed Faber and Staedler, "we have true simplicity in means of expression …. The maturing of this technique led to a new school of drawing." By means of its facility, the modern pencil lent itself to drawing performances that embodied new forms of looking and engaging the world outdoors and directly. For instance, Meder cites the German artist Adrian Ludwig Richter (1803–84), who fondly but astutely recollected that,

> with the new graphic means available in the early nineteenth century, we paid more attention to drawing than to painting. The pencil could not be hard enough or sharp enough to draw the outline firmly and definitely to the very

Figure 6.1 No. 322. John Herschel, *Glacier of Zermatt with the summit* (1) ascended on September 7. 20.1 × 29.9 cm. (From Larry J. Schaaf, *Tracings of Light: Sir John Herschel & the Camera Lucida* [1989])

> last detail. Bent over a paintbox no bigger than a small sheet of paper, each sought to execute with minute diligence what he saw before him. We lost ourselves in every blade of grass, every ornamental twig, and wanted to let no part of what attracted us escape . . . . [I]n short, each was determined to set down everything with the utmost objectivity, as it were in a mirror.[15]

Conducive to the capture of minutiae, bit-by-bit, these improved pencils allowed artists to lose themselves in detail as they carefully attended to the world with the pointed tip of their drawing utensils. It slowed the act of drawing to a form of attentive inspection of particulars. We may include artists like Jean-Auguste-Dominique Ingres (1780–1867), his student Théodore Chassériau (1819–56), Johann Carl August Richter (1785–1853), Jakob Friedrich Peipers (1805–78), and even Camille Pissarro (1830–1903) as examples of this new approach, which Herschel was very much a part. Indeed, Hershel's drawing productions must be seen as part and parcel of this new school.

Corot was also among the early enthusiasts of the pencil and "usually drew . . . so hard and sharp . . . it could leave a small trench in the paper. With this exacting instrument, he meticulously defined each tree branch and rock."[16] The indelible marks left behind by the modern pencil could no longer be simply erased using older methods like breadcrumbs but demanded the use of the recently introduced "rubber" eraser (first produced and sold as solid rubber cubes by the London mathematical, optical, and philosophical instrument maker Edward Nairne [1726–1806]). This is why Meder also stressed the essential role of much tougher "wove-paper" for this new school of drawing.[17] This innovative paper, which was even-surfaced, sturdier, and smoother than its "laid-paper" predecessor, was first manufactured by James Whatman at Turkey Mill in the 1750s and would come to be widely used by nineteenth-century artists and scientists, even recommended for William Fox Talbot's Calotype photographic process in the 1840s. Herschel too preferred Whatman's brand of paper – which remained the highest quality wove-paper for most of the century – for hundreds of his outdoor pencil drawings.

Historians have argued that drawing came of age as a medium just as these developments took hold among artists and scientists like Herschel.[18] Previous centuries had deemed the act of drawing as merely the means of producing undeveloped preliminaries, preparatory marks on their way to becoming some genuine artistic product such as a painting or sculpture. Thus, drawing was seen as always in need of completion, always remaining mere process. But the amalgam of novel drawing instruments, materials, and interests at the end of the eighteenth century reflected a new status for drawing as an end-result in its own right, filled with its own significance and meaning. Far from rough, careless, tentative strokes resulting in essentially disposable or recyclable sketches at the service of ostensibly superior forms of artistic expression, draftsmen like Herschel sought to attentively and deliberately depict nature with the pencil in hand in ways meant to endure as stand-alone and complete visual products (even if subsequently copied, colored, or printed). Much like the word "observation," "drawing" came to mean *both* process and product; both the means of attaining the record *and* the record itself. As a part of this broader historical shift in visual culture, Herschel's sketches should be treated as works in own their right, even if the bulk of them were never published in his lifetime or remained seemingly "incomplete." Tempting as it may be to dismiss hundreds of his drawings as subjective, tentative, or lacking in some way, Herschel's own conception of the practice of drawing must be better appreciated to understand how observation was embodied.

Camera Lucida Drawings

Perhaps no drawing instrument in Herschel's repertoire was more important than the optical device known as the camera lucida.[19] This was a portable, pocket-sized instrument developed by William Hyde Wollaston in around 1786, patented in 1806, and publicized as "An instrument whereby any person may draw in perspective, or may copy or reduce any print or drawing" – all ways in which Herschel would go on to subsequently use it.[20] With an adjustable brass stem whose bottom end could be clamped down to a drawing board or table, Wollaston's model employed a small quadrilateral glass prism attached to the top of the device. Looking down through the prism at just the right angle and direction, one's eye picks up a reflection of the scene in front of the prism, whose image then must be aligned with the image of the paper and pencil below so that they are brought into the same apparent plane, thereby transposing the scene onto the paper surface, ready to be drawn. The virtual image that seems to appear on paper due to this alignment – an image privy only to the user because it is technically not a projection at all but an illusion – depends for its maintenance on an absolutely steady drawing board but also the rigid placement of the draftsperson's head and body (see Figure 6.2). Considering that draftsmen like Herschel spent hours, usually out-doors, drawing in this rigid manner only hints at one of its many difficulties in practice. Once the alignment is achieved and maintained, the user graphically traces or "translates" the virtual image onto paper, preserving perspective, form, proportion, and orientation; all, unlike the camera obscura, without distortion. As Herschel's drawings dem-onstrate so well, the graphical results are striking for their clarity and quality of lines, each deftly placed with confidence and authority.

Although William Herschel had been one of the first to procure Wollaston's invention in February 1807 (for £2.12.6),[21] and John had become personally acquainted with Wollaston in 1814 while living in London, John's earliest known use of the camera lucida is from a drawing made in August 1816 and labelled "A cave in the cliff on the beach,

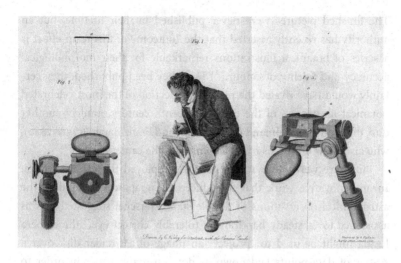

Figure 6.2 Frontispiece to *Description of the camera lucida: an instrument for drawing in true perspective, and for copying, reducing, or enlarging other drawings: to which is added ... A letter on the use of the camera by Capt. Basil Hall* (Sold by G. Dollond, optician to His Majesty, 59, St. Paul's Church Yard, London, who was the sole manufacturer of this instrument to the patentee, the late Dr. W. H. Wollaston [1830])

Dawlish, Devon" (No. 594). Done with some skill, the drawing, nonetheless, shows signs of being produced by someone who is a neophyte with the instrument. This observation has led Larry Schaaf to conclude that Herschel was most likely introduced to the hands-on *use* of the camera lucida in the summer of 1816 at the Dawlish home of his father's friend Sir William Watson, a notable draftsman with whom Herschel spent the summer days experimenting with optical instruments.[22] Be that as it may, from this point on Herschel assiduously employed the camera lucida for the rest of his life – a fact confirmed by the *General List.*

Herschel's camera lucida drawings included penciled landscapes and ruins, architectural and panoramic city views, and detailed drawings of geological formations. He engaged the camera lucida copiously at the Cape of Good Hope to draw, among other things, 132 drawings of some 200 to 250 species of indigenous bulbs he collected with his children and which they planted in their garden at Feldhausen.[23] Using the camera lucida, Herschel sketched the details of each specimen, subsequently colored with a sumptuous array of watercolors by his wife Margaret.

The finished pictures were never published in their lifetime, but an authority has recently asserted that the "outcome of this team effort is a series of botanical illustrations remarkable for their morphological accuracy and fidelity of colour."[24] Had they been published, they certainly would have rivaled the pictorial collection of the most celebrated botanical illustrators of the early nineteenth century, namely another duo, the brothers Ferdinand (1760–1826) and Franz Bauer (1758–1840), who also drew hundreds of specimens with the camera lucida.[25]

But there were other ways Herschel employed the camera lucida. In an innovative work that introduced the graphical method to the astronomical observations of double stars, Herschel suggested that as assistance to "a steady hand and a tolerably correct eye," the camera lucida might be used to "project" an image of a circular disk over a scatter of data-points laid down as dots on graph paper in order to achieve, after some manipulation of the sheet of paper and optical instrument, "a tolerably correct ellipse." Doing so with a "moderately-hard pencil and smooth paper without water-marks" could reveal the probable orbit of one star around another in a binary system.[26] Such instrumental assistance supported the critical and contested issue of literally drawing extrapolations (as lines or curves, those "large and graceful sinuosity") by means of judgment rather than calculation.[27] Herschel used the camera lucida as a kind of projection in other ways as well. He employed it, as we shall see, to patch together different parts of multiple, separate drawings of one and the same nebula. He would do something similar for his panoramas, which combined different views through the camera lucida aided with a specially made rotating drawing table.[28] It was also with his trusty camera lucida that Herschel made copies of his own drawings, for instance, gifting Charles Babbage's wife Georgiana with a number of "enlarged" (meaning enriched) presentation copies in 1821 of drawings he had made on one of his Grand Tours with Babbage (see Figure 6.3).[29]

Indeed, using the camera lucida to copy drawings was a well-known practice among nineteenth-century engravers and lithographers. As both Herschel and the engravers knew, the device could also be used to resize the original drawing. This is how, for example, John James Audubon (1785–1851) reduced the size of his original lavish double-elephant-size folio edition (1827–38) of The Birds of America into the first octavo edition (1839–44). His eldest son Victor Gifford

Figure 6.3 No. 517 "enlarged copy . . . made for Mrs. Babbage," from No. 322 *Glacier of Zermatt with the summit.* 21.0 × 30.7 cm. (From Larry J. Schaaf, *Tracings of Light: Sir John Herschel & the Camera Lucida* [1989])

Audubon would later use the camera lucida again to resize the latter edition even further by one-eighth into an even smaller and thus more affordable edition (1856).

As with the case of the modern pencil and new forms of paper, Herschel was only one of many to utilize the camera lucida. After its invention, the device is said to have "spread epidemically."[30] Just nine months after sales of Wollaston's camera lucida had begun by George Dollond (1774–1852) in London, and despite its prohibitive price, an estimated 1,000 instruments were sold.[31] It was thus probably much more widely used than publicly acknowledged by artists. It has been argued that one of the masters of the modern pencil, Ingres, regularly used the instrument; as did Sir Francis Chantery (1781–1841), who incorporated it into his factory-like process of making sculptures. The Danish painter Christen Købke (1810–48) deployed it to some effect, just as Gustave Caillebotte (1848–94) did much later.

Scientifically inclined users, on the other hand, were less hesitant to openly declare its utilization; in fact, its use was sometimes publicly declared as a reassurance to readers that illustrations were reliable and exact. Here we may include the ethnographic drawings by Captain Basil Hall (1788–1844) of North America (1829); the ethno-archaeological depictions of the Mayan ruins by Frederick Catherwood (1799–1854) in 1841 and 1844; Sir William Gell's (1777–1836) celebrated engravings of Pompeii (1818); some of the drawings by the astronomer Charles Piazzi Smyth (1819–1900); and of course Robert Hay (1799–1863), dubbed the "patron saint" of the camera lucida, who funded an expedition to Egypt and hired a band of avid camera lucida users for the tour, including Joseph Bonomi the younger (1796–1878), the architect Francis Arundale (1807–53), the orientalist Edward William Lane (1801–76), and the antiquarian George Alexander Hoskins (1802–63).[32] If one includes all the microscopists who began to systematically use the camera lucida attached to the microscope to draw what they saw through it – especially after the Italian optician and instrument maker Giovanni Battista Amici's clever modifications to Wollaston's model – one gets a sense of how widespread its use was in both the arts and sciences. This is all to say that Herschel, who would also come to prefer Amici's version of the instrument, was part of a much wider European-wide tendency; and yet, as Schaaf correctly points out, he "was to produce more camera lucida drawings – of higher quality and interest – than any other known individual."[33]

In contrast to some of the aforementioned camera lucida users who favored wash instead of pencil to indicate tone, who started with shadows before hard outlines, or worked over pencil lines with India ink, Herschel preferred not to apply any of these techniques, relying on the variations of his pencils in hardness and darkness instead. But one further notable feature set Herschel's practice apart from contemporaries. This stemmed from the widely recognized quality of the camera lucida to afford wide fields of view without distortion to perspective and scale, thereby showing things as they might appear to the naked eye when looking at the same scene. In several of his drawings, Herschel exploited these qualities in order to measure and preserve the relative metric of the scene. In addition to including the cardinal directions, Herschel sometimes incorporated axes X, Y, and Z, while in others he labeled conspicuous features in the landscape (e.g., mountain tops or buildings) with a, b, c, etc.[34] In some cases, the latter are determined in inches by the coordinates of X, Y, and Z (e.g., No. 398). In other cases,

Figure 6.4 No. 298 John Herschel, *Valley of Fassa from Opposite Gries* (August 1824). The J. Paul Getty Museum, Los Angeles, Gift of the Graham and Susan Nash Collection, 91GG.98.32. Image courtesy of Getty's Open Content Program. Note Herschel's annotations throughout the drawing, including notes on lighting, thickness of trees, location of dolomite rock deposits, etc.

Herschel measured the angles of ab, bc, ac in degrees, minutes, and seconds (see Figure 6.4).

Both are tethered to a control point sometimes indicated in the drawings by the symbol "⊙" that represents the position of the prism above the paper and thus the size of the projected scene on paper. Along with Z, which gives the height of the eye in inches from the surface of the paper, Herschel figured several crucial relationships internal to the scene. Among other things, this allowed Herschel, at least in principle, to determine not just the relative position and order of the parts and details found therein, but also to re-identify the exact location wherefrom the scene was acquired and thus potentially to take it again.[35] All together, these observations could very well contribute to the tracking of change in, say, topography or geological features of a landscape (e.g., Table Mountain).[36] They also represent a particular attitude concerning the preservation of scale and metric that we will also find in some of Herschel's drawings of the nebulae.

Refined Observation

The camera lucida allowed the user to draw with extraordinarily accuracy. Yet it was no royal road to drawing, as Henry Fox Talbot (1800–77) had famously made clear.[37] Besides requiring much practice, the camera lucida did not simply replace the requisite skills required for good draftsmanship as much as it *enhanced* them.[38] Moreover, the instrument heightened the observer's experience of the world, making it conducive to a particular kind of observation. Schaaf describes the camera lucida as encouraging a "tightly analytic observation" of the world.[39] He calls it "a contemplative instrument," by which he means that the camera lucida "demands a detailed analysis of the component details of a scene," and as such, "facilitated the gathering of visual facts."[40] Brian Warner has claimed that in using the camera lucida, "one of [Herschel's] skills was an ability to indicate the texture of foliage so well that it was often possible immediately to identify the species of tree."[41] There can be no doubt that the level of detail permitted by the camera lucida, particularly in conjunction with the range of marks made possible by the modern pencil, was extraordinary. Together these drawing instruments parsed the world for close analytic examination of particulars. There is little here, to be sure, that echoes the sage generalizations – be they ideals, characteristics, or essences – of a previous generation and their epistemic virtue of "truth-to-nature."[42]

Yet instead of losing himself in the minutiae of a blade of grass, as other pencil enthusiasts of the "new school of drawing" may have, Herschel's practice demanded an active mind permanently vigilant on at least two interrelated fronts: beauty and harmony, on the one hand, and relations and synthesis, on the other.[43] Herschel writes in the *Discourse* that the observer "contemplates the world, and the objects around him, not with a passive, indifferent gaze, as a set of phenomena . . . but as a system disposed with order and design. He . . . feels the highest admiration for the harmony of its parts."[44] Herschel sought a productive back-and-forth between analysis and synthesis. Elsewhere, when advancing what he called the *picturesque*, Herschel makes it abundantly clear that the observer must accommodate both, "embracing with distinctness and truth a sufficiently extensive view . . . [while] . . . duly supressing detail. Such a view of nature," he continues, ought to be, in the highest possible sense of the word, *picturesque*, nothing standing in relation to itself alone, but all to the general effect. In such a picture every object is suggestive . . .. Nature, indeed, offers all in her profusion, and complete in all its details; and the contemplative mind finds among them paths for all its wanderings, harmonies for all its moods. But such exuberance [for detail] is neither attainable nor to be aimed at in a descriptive outline, where leading features only have to be seized, which imagination is stimulated to fill up by the grandeur of the forms, and the intelligible order of their grouping.[45]

If the productive and harmonious amalgam of detail with its ordered setting was sought for the sake of science, it seems Herschel utilized the camera lucida not so much for detail – as he might have done with pencil and paper alone – but to help him situate that detail into some significant system of relations and composition suggestive, for example, of expected actions, motions, or behaviors to come. Though the optical device was "well known as corrective of the . . . decisions of the eye or a succedaneum in the labour of educating that organ,"[46] it did not achieve the back and forth between analysis and synthesis automatically or mechanically; rather, it required operation by someone with sound judgment and an "attention more or less awake."[47] This is no instance of "mechanical objectivity."

What is called for on the part of the observer is an informed discrimination of "leading features," which picks out only those details and relations that when taken together suggest something beyond themselves, culminating in some "law of nature" rather than a mere "historical event." Though each of the components of the picturesque are associated and suggestive of one another, the correct and most effective relation can only be achieved through attention and judgment in the process of recording or translating an observation. Or as Herschel put it, again in the *Discourse*: "the circumstances . . . which accompany any observed fact, are main features in its observations, at least until it is ascertained by sufficient experience what circumstances have nothing to do with it, and might therefore have been left unobserved without sacrificing the fact." When the circumstances are utterly novel, the observer "ought not to omit any circumstance capable of being recorded." In either case, as surmised above, "our observation of external nature is limited to the mutual action of material objects on one another; and to facts, that is, the association of phenomena or appearances."[48] The "perfect observer" will thus have "his eyes as it were opened," looking upon the world with "the eye of the philosopher."[49]

Erna Fiorentini has called this form of observation "refined observation." She explains that in contrast to "exact" observation, in this form "sensory experience, attention and curiosity of the observer are expected to act in concert with elaborate instrumental equipment, in a process of mutual assistance, which is understood as the only way to attain the most accurate results." She argues that the camera lucida "perfectly matches the logic of 'refined observation' revealing itself in the *Discourse*, namely the adjustment of subjective impression and objective data collection via discernment."[50] According to Fiorentini, while using the instrument,

> the mental activity directed to the technical accomplishment of the drawing ... recedes in favour of an inspection of the observed in terms of its meaning. This kind of inspection comprises the analysis and the estimation of the perceived data, and requires at the same time the selection of those particulars deemed relevant to accomplish the goal of the representation. That means that instead of concentrating on the technique to adopt for a correct transposition of the observed scene, the Camera Lucida user could focus his or her attention to consider the relevance of its whole and of its singular components .... This ongoing process of 'perceptual and intellectual possession' induced while seeing with the prism is the most important peculiarity of observing and recording with the Camera Lucida.[51]

When it comes to drawing, these statements imply, and Herschel's practice shows, that the act of drawing is more than *only* a record but also a systematic process of close study, inspection, and reflection, where judgment is crucial for the assessment and relevance of what is included or excluded from the scene drawn. If the act of drawing is enhanced by the use of the camera lucida, we may regard it as more than just a recording device but rather one that heightens scientific or refined observation in the sense of the picturesque.

Yet it is precisely here that Herschel's own drawing practices, in the service of a refined observation, part company with his own account of observation in the *Discourse* when he proclaims: "Recorded observation consists of two distinct parts: 1st, an exact notice of the thing observed, and of all the particulars which may be supposed to have any natural connection with it; and, 2ndly, a true and faithful record of them." In other words, the determination of sensations into some distinct perception comes first and then follows a faithful record of it. Herschel's own observational practices suggested, however, the collapse (or at least partial overlap) of these two distinct parts. This is more than drawings containing things unnoticed *in situ*, lying dormant only to strike one upon further examination *ex post facto*, in some "after-rumination."[52] The effect is more profound: one sometimes gets clearer about what one is observing only in the course of producing a true record of it; or, in the process of determining what one actually perceives by means of painstakingly piecemeal drawings with a camera lucida, one is ultimately also recording. Observational drawing practices, then, are both a record and a means to closely attend and decide what one actually sees.[53] With "respect to our record of observations," wrote Herschel, "it should be not only circumstantial but *faithful*; by which we mean, that it should contain all we did *observe*, and nothing else."[54] The camera lucida was an instrumental means of enhancing both components of observation: product and process – and often at the same time.

Lest we forget how difficult it is to observe something even in front of one's own eyes, and thus how important it is to be aided in the process of observation, it is instructive to recount the time Herschel trained Charles Babbage to see the dark lines of a solar spectrum. "A striking illustration," reported Babbage,

> of the fact that an object is frequently not seen, from not knowing how to see it, rather than from any defect in the organ of vision, occurred to me some years since, when on a visit at Slough. Conversing with Mr. Herschel on the dark lines seen in the solar spectrum by Fraunhofer, he inquired whether I had seen them; and on my replying in the negative, and expressing a great desire to see them, he mentioned the extreme difficulty he had had, even with Fraunhofer's description in his hand and the long-time which it had cost him in detecting them. My friend [Herschel] then added, 'I will prepare the apparatus, and put you in such a position that they shall be visible, and yet you shall look for them and not find them: after which, while you remain in the same position, I will instruct you how to see them, and you shall see them, and not merely wonder you did not see them before, but you shall find it impossible to look at the spectrum without seeing them.'[55]

More often than not, that is, nature challenges the observer, who may have grown accustomed to perceiving it in one way, to gain leverage over it in order to reveal those features otherwise overlooked. In some cases, the leverage is an instrument like the camera lucida. If spectral lines posed such difficulties, imagine perceptually confronting not a series of fine lines but a whole landscape with complex geological structures never before seen, or heretofore unknown architectural styles. James Fergusson (1808–86), for example, encountered architectural wonders in India that he exclaimed were "quite impossible for any artist who never saw a building of that class . . . and has no knowledge of the style himself, nor any means, from real specimens of acquiring it, to render that peculiar character or physiognomy which the style possess."[56] It was only with the aid of the camera lucida that Fergusson achieved the "impossible," namely, his *Picturesque Illustrations of Ancient Architecture in Hindostan* (1848).

Refined Observations of Celestial Objects

There is no more remarkable example of this productive collapse between exact notice and faithful record than the observation of nebulae, sidereal objects that were almost entirely enigmatic to nineteenth-century astronomers. Like his father before him, John Herschel was one of the foremost investigators of these baffling objects. On presenting the Royal Astronomical Society's prestigious gold medal to Herschel for his 1833 catalog of nebulae and clusters of stars, which also contained ninety engraved plates made after pencil drawings by Herschel's hand, George Airy (1801–92) made it abundantly clear that such phenomena, in fact, defy words and "numerical expression." "So, let no one suppose," Airy proclaimed, that "I am overrating the value of these drawings . . .. Few observers possess the delicacy of hand of Sir John Hershel; yet it were to be wished that his example might be imitated by many, and that careful drawings, the best that circumstances admit of, might frequently be made of the same nebula."[57] Herschel, to be sure, was not just cataloging thousands of nebulae and clusters; he was also making delineations of many in a separate series of "pointed observations."[58] In fact, the over 150 finished pencil drawings of nebulae and clusters engraved and printed for his official publications on them between 1826 and 1847 exemplify Herschel's most significant use of drawing for the purposes of observation in science.

Considering the sheer difficulty in observing such *barely visible* and complex objects at the telescope, coupled with the fact that it was unclear what astronomers were looking at or how these objects were to be described, the act of drawing became one of Herschel's main lines of attack in his observational approach to nebulae and clusters.[59] Any additional instrumental means, in conjunction with the largest tele- scopes in the world, that could aid in achieving the right kind of drawing for "refined observations" was thus welcome. Such means must assist the observer-draftsperson to situate details of the object in the context of a system of relations, slow down the observer's gradual approach to the phenomena, and aid in making sense of what was seen. Moreover, these nebulae and clusters raised questions of change and development, important for determining the "laws of nature" that might govern them – precisely the purview of refined observations that purported to follow "leading features" in context of the picturesque (see Chapter 3). Especially when taken together as a series, drawings – particularly those designed to preserve relations and context as well as details – would aid in the detection of changes in these objects, indicat- ing the direction, frequency, and intensity of such changes. It is no wonder that Airy, like Herschel, made a plea for more and better drawings. Wood-encased pencils, wove-paper, and a propensity to draw outdoors were all part of Herschel's repertoire when he drew clusters and nebulae at the eyepiece of the telescope; yet here the camera lucida was unavailable to him.

The trouble was the camera lucida could not be employed to much effect at the eyepiece of a large reflecting telescope like the one Herschel used. One issue was that of illumination: paper surface and target object were too faint. This is not to say that attempts to combine the camera lucida with the telescope were not attempted. Already in 1809, R. B. Bate (1782–1847), an early advocate of the camera lucida, recom- mended its use with a microscope or telescope.[60] In 1811, Cornelius Varley (1781–1873) came up with what he called the Graphical Telescope, in which a camera lucida was integrated into a low powered telescope with an object glass.[61] Varley successfully used this bulky instrument to produce a number of drawings, mostly landscapes, but with a weak telescope it had no significant astronomical application.[62] Interestingly, in 1836, a summary of a letter of Charles Babbage to Francis Baily appeared in the *Monthly Notices of the Royal Astronomical Society* that suggested, "the application of a camera lucida to a telescope mounted equatorially, and governed by a clock; the images of groups of stars or of spots on the moons disc, to be thrown on varnished plate of copper, and traced with a steel point: the maps thus obtained to be etched. Mr. Babbage thinks, from some experiments formerly made by him, that there would be no great difficulty, and hopes that the attempt will be made by some person possessing means and leisure."[63] It is hard to tell if anything came of these experiments, but Herschel, who at this time was at the Cape of Good Hope, was employing another scheme altogether, one based on a system of relations produced by fundamental stars plotted not on a copper plate but on paper with pencil, straight-edge, and compass. The camera lucida, it turned out, was not the only means to refined observations with pencil and paper.

If not with a camera lucida, how did Herschel have "his eyes opened" at the telescope with the "the eyes of the philosopher" while observing nebulae and clusters? Herschel formulated another sort of instrumental means, not of brass and glass but of paper elaborately prepared to control the reception of the details of nebulae by delicate pencil work: in particular, the painstaking manner of work that went into the production of eight out of fifty-eight drawings engraved and printed for the *Cape Results* of 1847.[64] A full third of the four years of nightly observations at the Cape of Good Hope (1834–38) were dedicated to drawing these objects at the eyepiece of his large twenty-foot (focal length) reflector, which he had brought with him from England. For some of the more delicate measurements, Herschel used his much smaller and more precise seven-foot (focal length) equatorial telescope, which was fitted with a precision micrometer. Upon his return home, Herschel continued filling in, mapping, calculating, reducing, and cataloging these objects for at least another six years. The elaborate depictions of nebulae and clusters in the *Cape Results* represent Herschel's response to the dissatisfaction he felt with his own drawings made previously with the same telescopes in England, including those from 1833 – the same drawings that Airy had previously praised. In other words, the *Cape Results* drawings exemplify what Herschel thought to be the most advanced means of visually representing nebulae. I have referred to these elsewhere as *descriptive maps* in contrast to Herschel's less detailed drawings, which I have dubbed *portraits*.[65] What made these descriptive maps superior to any that came before was the observational procedures employed in drawing them.

Unlike the earlier series of nebulae observations made in England, Herschel executed these new drawings under the wonderfully clear and stable skies of the Cape of Good Hope. This meant, in the words of Herschel, that "the beauty and tranquillity of the climate is such, that the stars are reduced to all but mathematical points, and thus allow of their being viewed like objects under a microscope."[66] Herschel's Cape drawings exploited these pristine viewing conditions to carefully measure stars he plotted onto paper to preserve their relative positions. This groundwork afforded a stepwise series of inclusions, over a number of years, of more and more graphical, pictorial, and numeral information penciled into these maps in ways that preserved the metric of relations alongside the intricacies of physical features and details of the nebulae. Yet the engraved plates in the *Cape Results* show little signs of this backstage work; rather, they appear as though completed in a night or two (see Figure 6.5). We must, therefore, turn to Herschel's archive, where we find "mono-graphs": dedicated folders for each of the eight descriptive maps in the *Cape Results* that contain significant amounts of paperwork, including series of what Herschel called "working skeletons" that formed the substratum for these meticulous drawings.[67]

Due to the extreme delicacy of making measurements and seeing the faint stars involved in and around the nebulae, and because of the many possible sources of error in such observations, Herschel stressed that "such figures in fact cannot be adequately described and figured in a single night." Only by taking the results of many nights of observations could an observer avoid or at least decrease the errors involved. This meant not only many repeated measurements but also repeated draw-ings night after night. To achieve this, Herschel suggested "breaking up [the paper-surface] into triangles [each] to be explored in detail."[68] Inspired by land-surveying and cartographical techniques, Herschel's chains of triangles are the "working skeletons" utilized to triangulate

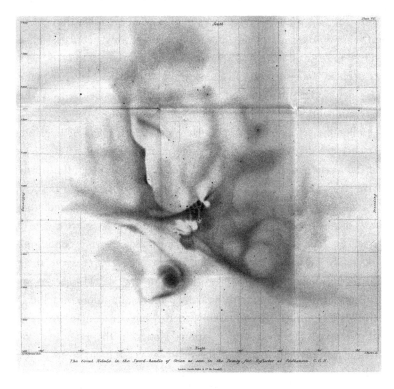

Figure 6.5 Engraved drawing of "The Great Nebula in the Sword-handle of Orion as seen in the Twenty-feet Reflector at Feldhausen," in John Herschel's *Cape Results* (1847), plate VIII

the approximate, relative positions of stars and the meticulous entry of nebulous material on to paper.

Herschel constructed his working skeletons in a series of steps. The first was to use the micrometer of the smaller seven-foot telescope to determine the relative position-angles and angular distances of the more conspicuous stars in and around the nebula. From these Class 1 stars, a "zero star" was selected as near the center of the nebulae as possible. This star formed the "zero point" of the "chief line" that centered a grid divided so that the x-axis plots Right Ascension (RA) and the y-axis North Polar Distance (NPD). With this grid in place, other directly measured stars (Class 1 stars) were inserted into their coordinated positions. Once the Class 1 stars were in position, Herschel used the base lines they formed to triangulate the second class of stars.

Class 2 stars are those whose coordinates are determined not by direct measurement at the telescope but by using the network of triangles whose initial base is made up of two Class 1 stars. Herschel went as far as to regard the mean position and distance of each Class 2 star determined in this way as having "a degree of exactness not inferior to what would have been afforded by direct measurements with the position micrometer."[69]

From these triangulations, chains of triangles were generated connecting Class 1 and Class 2 stars and covering the entire area of the nebula and surrounding regions. Once the chain of triangles had been sufficiently spread over the area of the nebula – which had yet to be drawn in – stars much too faint to be directly measured were inserted *by estimation*. These were Herschel's Class 3 stars, and their mean coordinates in RA and NPD were determined in relation to everything already laid down in the working skeleton. This sort of entry was the result of "mental comparison,"[70] wrote Herschel, and involved *judgment* on the part of the observer, controlled and enhanced by the totality of what had already been laid down as a system of relations. Finally, all classes of stars had their mean coordinates reduced and entered into a catalog of stars. (The catalog of stars derived in this way for the Eta Argus nebula numbered over 1,200 individual stars.)

Once laid out, working skeletons organized and controlled the subsequent reception of the complexities of a nebula by hand at the telescope. Using the twenty-foot reflector, Herschel methodically sketched in with pencil, little by little, the nebulous body with all its details into the already-constructed skeleton.

The skeletons operated to coordinate hand and eye so that the endless details of the nebulae could be precisely situated. The established skeleton thus controlled how and where a nebula was subsequently incorporated into a system of relations, steadily accumulating over time and many nights. The skeletons' triangles forced Herschel to examine each in turn, creating an opportunity for systematic entry of even more stars and nebulosity by pencil into each individual triangle. The variously irradiated wisps, strata, branches, appendages, and cloudy haze of an ill-defined nebula were all gradually "laid down within each triangle by the sole judgment of the eye, [for] it is not possible to eliminate the errors of such judgment by any system of calculation," explained Herschel.[71] Though procedural, the production of descriptive maps was not mechanical.

For more complex nebulae, the same network of triangles was either filled in more than once or many working skeletons were constructed for the same object. (So, for example, there are up to twenty-three individual working skeletons in the monograph for Eta Argus alone.) In either case, working skeletons were used again and again, on many different nights and over years to insert and reinsert, regulate and adjust what was seen of nebulosity and the stars that grounded it. The use of working-skeletons, wrote Herschel, "is the only mode in which correct monographs can be executed of nebulae of this kind, which consist of complicated windings and ill-defined members obliterated by the smallest illumination of the field of view." The level of detail, which could gradually and precisely be entered between the lines of the triangles, rooting it into a well-ordered and congruous scene, was extraordinary. It seems to have worked so well to focus Herschel that he, quite uncharacteristically, emoted that "while working at the telescope on these skeletons, a sensation of despair would arise of ever being able to transfer to paper, with even tolerable correctness, their endless details."[72] His only safeguard were the skeletons themselves, which counteracted the immense amount of detail to be sketched, compelling Herschel to systematically reorient himself in relation to the whole. Thanks to these artificial means, the niceties of the nebula or cluster could be entered harmoniously among its other parts and thus as a whole – exactly what a refined observation of the picturesque called for.

Near the conclusion of this extensive drawing process, Herschel would have accumulated several separate sheets with skeletons upon each for the same object. These then had to be combined into *one image* of the nebula on a single sheet to form its standard descriptive map, then prepared and readied for the engraver's plate and thus printing and publication. On the one hand, thanks to their carefully measured and preserved scale, the working skeletons themselves assisted in the transfer of star positions from one sheet to another. This was especially helpful in cases where Herschel had found the best configuration of chains of triangles and wanted to preserve them throughout the succeeding series of skeletons. On the other hand, there are indications that Herschel used the camera lucida to patch together different working skeletons, each sheet containing a drawing of some distinct part of the same nebula, into one congruent and whole image. Aligning each successive sheet with one master copy, Herschel likely used the camera lucida to enter not just stars in a measured way but the nebulosity from the original sheets and into the copy, which gathered all the portions of the nebula into one place in a coherent manner. I believe this is how he constructed the finished descriptive maps for the more complex nebulae, particularly M42 and Eta Argus. The *General List* of all his drawings includes these two maps numbered as No. 726 and 727 respectively, with the accompanying note that they were framed, and doubtless hung at his home. But the finished drawings were also meant to be engraved for the *Cape Results*, each of which contain no indication of the many skeletons that formed their substratum but only a nebula or cluster superimposed on the faint grid behind them.

A continuity of information was thereby maintained between multiple skeletons made for one object over many years. Even when Herschel was back in England and busy preparing his findings for publication, the skeletons afforded continued measurements and calculations, judgment and drawing, even away from the telescope. They provided consistency, regularity, and scale to Herschel's descriptive maps of deep-sky objects notoriously difficult to see and measure. But above all, the skeletons coordinated and focused what the eye saw and what the hand entered onto paper into a system of relations that permitted Herschel to scrutinize each part of a nebula in a methodical and highly attentive manner. By means of these observational procedures, which dramatically slowed both the act of drawing and seeing, Herschel became intimately familiar with such demanding objects and their endless details. The resulting published descriptive maps of nebulae furnish an excellent example of the collapse between making a record and deciding what is perceived.

Conclusion

When eventually confronted with the question of photography, hand-drawn descriptive maps would go on to inform Herschel's expectations of what he considered a properly scientific photographic presentation of nebulae.[73] This does not mean that descriptive maps somehow anticipated the results of photography or were a precursor to it. It is only to say that the kind of preparation and display of information expected of proper scientific observations was directly related to the production and function of descriptive maps. Thus, in a published 1853 letter about the aims of a newly conceived southern telescope in Melbourne, Australia, Herschel advised that when it came to nebulae, photography could only be effectively used to "impress on paper a skeleton picture [of] the images of *the stars only* which might accompany or be disseminated over a given nebula to be delineated [by hand]."[74]

Given photography's technical limitations in this period, Herschel was endorsing an abbreviated role for photography, one in the service of drawing: that astronomers take photographs of the brighter stars in or around a nebula, instead of the nebula itself, to act as the paper basis for the drawing-in of nebulous material by hand.[75] Herschel's proposal not only derived from his own procedures of *drawing* nebulae at the Cape of Good Hope, well before he knew anything of the invention of photography in early 1839, it also echoed Babbage's experiments mentioned above with the camera lucida for the purposes of astronomical observations. Both approaches involved the careful placement of stars as groundwork. This points to an important common feature of depicting with camera lucida, working skeletons, and photography: each laid down parameters for a system of relations governing the orderly entry of parts into a greater whole that could be scaled and measured.

Herschel referred to fundamental stars (e.g., Class 1 stars), whether drawn, etched, or photographed, as "established authentic landmarks."[76] As early as 1827, Herschel reminded his audience of the importance of having a good list of astronomical landmarks or "zero points" that would guide ships, calibrate instruments, correct clocks, and aid in measurements of the heavens and their subsequent reductions. The stars, according to Herschel, are the landmarks of the universe which "teach us to direct our actions by reference to what is immutable in God's work."[77] Herschel's observational practices are instances of his hands, eyes, and mind being steered in reference "to what is immutable." Even though some free-hand drawings exist in his extensive oeuvre (designated as eye sketch or draft), when it came to scientific observations Herschel preferred drawing practices that were aided with some instrument, whether camera lucida or a working skeleton. With instrumental aide to proactively constrain how he drew, Herschel aimed at generating productive tension between analysis and synthesis for refined observation. Though procedural, this form of observation was not mechanical; rather, judgment, estimation, and discernment – indeed an awakened mind – were essential.[78]

Herschel's observations of diverse phenomena were disciplined by the very act of drawing that constrained and guided them and the judgments they incorporated. These were carefully planned and able performances that coupled acts of seeing and drawing in ways that were conducive to a range of observational sciences. Herschel's refined observations, therefore, are just one instance in a much longer history of what counted as observation in the sciences, an important instance that exemplifies both critical overlaps of science and art – not so much at the level of results but of processes and tools – and an integration of record and attention. In this way, Herschel's refined observations are a robust example of what historians have identified as "paper knowledge" or "paperwork," challenging the same historians to include, in addition to their overriding focus on text and number, the neglected role of sketch-making.[79] Yet Herschel's paperwork contributed directly not just to observation as memory or record but to attending and looking closely. Herschel's drawings are indeed "paper tools" that deserve another look.[80]

Notes

1 John Herschel, "Account of Some Observations Made with a 20-feet Reflecting Telescope . . .," *Memoirs of the Astronomical Society of London*, 2 (1826): 487–95, 488–89. For discussion, see O. W. Nasim, "Extending the Gaze: The Temporality of Astronomical Paperwork," *Science in Context*, 26 (2013): 247–77.
2 Charlotte Klonk, *Science and the Perception of Nature: British Landscape Art in the Late Eighteenth and Early Nineteenth Centuries* (New Haven: Yale University Press, 1996).

3 Michael Baxandall, *Painting and Experience in Fifteenth-Century Italy*, 2nd ed. (Oxford: Oxford University Press, 1988). But also see, for the history of science context, Klaus Hentschel, *Visual Cultures in Science and Technology: A Comparative History* (Oxford: Oxford University Press, 2014).

4 Ann Bermingham, *Learning to Draw: Studies in the Cultural History of a Polite and Useful Art* (New Haven: Yale University Press, 2000). See also Richard Sha, "The Power of the English Nineteenth-Century Visual and Verbal Sketch: Appropriation, Discipline, Mastery," *Nineteenth-Century Contexts*, 24 (2002): 73–100.

5 Larry J. Schaaf, "John Herschel, Photography and the Camera Lucida," *Transactions of the Royal Society of South Africa*, 49 (1994): 77–102, 77.

6 See Günther Buttmann, *The Shadow of the Telescope: A Biography of John Herschel* (Guildford: Lutterworth Press, 1970).

7 Quoted in Brian Warner, *Cape Landscapes: Sir John Herschel's Sketches 1834–1838* (Cape Town: University of Cape Town Press, 2006), 16. For more on Herschel at the Cape of Good Hope, see E. G. Musselman, "Swords into Ploughshares: John Herschel's Progressive View of Astronomical and Imperial Governance," *British Journal for the History of Science*, 31 (1998): 419–35.

8 See Larry J. Schaaf, *Tracings of Light: Sir John Herschel & the Camera Lucida* (San Francisco: Friends of Photography, 1989), 10.

9 John Herschel, "General List of all My Drawings and Sketches. Made in July 1861," Harry Ransom Humanities Research Center, The University of Texas at Austin (hereafter HRC). I have used the virtual copy available of this at http://vision.mpiwg-berlin.mpg.de, accessed February 15, 2023.

10 Herschel was also a collector of art works. The most famous example is the relatively large collection of Sir Joshua Reynolds drawings he owned. See Luke Herrmann, "The Drawings of Sir Joshua Reynolds in the Herschel Album," *The Burlington Magazine*, 110 (1968): 650–58.

11 Pierre-Henri de Valenciennes, *Elemens de perspective pratique à l'usage des artistes* (Paris, 1799).

12 Frauke Josenhans, "Sur le Motif: Painting in Nature around 1800," *Getty Research Journal*, 1 (2009): 179–90.

13 Quoted in Thomas Buser, *History of Drawing*, https://historyofdrawing .com/?page_id=9, accessed February 15, 2023.

14 Richard Taws, "Conté's Machines: Drawing, Atmosphere, Erasure," *Oxford Art Journal*, 39 (2016): 243–66, on 254. For more on the history of the pencil, see Philip Rawson, *Drawing: The Appreciation of the Arts* (Oxford: Oxford University Press, 1969); Henry Petroski, *The Pencil: A History of Design and Circumstances* (New York: Knopf, 1990); and Caroline Weaver, *The Pencil Perfect: The Untold Story of a Cultural Icon* (Berlin: Gestalten Verlag, 2017).

15 Joseph Meder, *The Mastery of Drawing*, Winslow Ames, trans., vol. 1 (New York: Abaris Books, 1978), 117–18.

16 Quoted in Buser, *History of Drawing*.

17 Joseph Priestley introduced the world to Nairne's solid rubber cubes made for the distinct purpose of erasing the marks left behind by "black-lead pencils" and claimed each could last several years. See his *A Familiar Introduction to the Theory and Practice of Perspective* (London, 1770), xv. As an instrument maker, Nairne also worked closely with astronomers; see, for instance, *An Observer of Observatories: The Journal of Thomas Bugge's Tour of Germany, Holland and England in 1777*, edited by K. M. Pedersen and Peter de Clerq.

18 Taws, "Conté's Machines," 254. Also see Deanna Petherbridge, *The Primacy of Drawing: Histories and Theories of Practice* (New Haven: Yale University Press, 2010); and David Rosand, *Drawing Acts: Studies in Graphic Expression and Representation* (Cambridge: Cambridge University Press, 2002).

19 For details, see John H. Hammond and Jill Austin, *The Camera Lucida in Art and Science* (Bristol: Adam Hilger, 1987).

20 William Hyde Wollaston, "Specification of the Patent granted to William Hyde Wollaston, of the Parish of St. Mary-le-Bone, in the County of Middlesex, Gentlemen; for an Instrument whereby any Person may Draw in Perspective, or may Copy or Reduce any Print or Drawing," *The Repertory of Arts, Manufactures and Agriculture*, X, Second Series LVII (1806).

21 Melvyn C. Usselman, *Pure Intelligence: The Life of William Hyde Wollaston* (Chicago: University of Chicago Press, 2015), 150.

22 Schaaf, *Tracing*, 15 and figure 12.

23 See Brian Warner and John Rourke, *Flora Herscheliana: Sir John and Lady Herschel at the Cape 1834 to 1838* (Houghton: The Brenthurst Press, 1996).

24 John Rourke, "John Herschel and the Cape Flora, 1834–1839," *Transactions of the Royal Society of South Africa*, 49 (1994): 71–86, 79.

25 H. Walter Lack, "Recording Form in Early Nineteenth Century Botanical Drawing: Ferdinand Bauer's 'Cameras,'" *Curtis's Botanical Magazine*, 15 (1998): 254–74.

26 John Herschel, "On the Investigation of the Orbits of Revolving Double Stars . . ." *Memoirs of the Royal Astronomical Society*, 5 (1833): 171–222, 183, 179.

27 Thomas Hankin, "A 'Large and Graceful Sinuosity': John Herschel's Graphical Method," *Isis*, 97 (2006): 605–33.

28 Warner, *Cape Landscapes*, 77.

29 Schaaf, *Tracings*, 37.

30 Erna Fiorentini, "Optical Instruments and Modes of Vision in Early Nineteenth Century," in Werner Busch (ed.), *Verfeinertes Sehen: Optik und Farbe im 18. und frühen 19. Jahrhundert* (Berlin: De Gruyter Oldenbourg, 2008), 201–21, 204.

31 Usselman, *Pure Intelligence*, 149.

32 Schaaf, *Tracings*, 30. See also Catherine Delano Major, "Illustrating, B.C. (Before Cameras)," *Archaeology*, 24 (1971): 44–51.

33 Schaaf, "John Herschel," 89.

34 See Herschel, "General List," 19; but also Schaaf, *Tracings*, 113.

35 These markings are often erased or exist very faintly, sometimes not even in the bounds of the picture but beyond its frame. Often one needs to really search for them in the picture, especially the control point, which is either erased or tends to disappear into the details. This is in part explained in Herschel, "General List of all My Drawings and Sketches. Made in July 1861," 4–5; but also in Schaaf, *Tracings*, 113.

36 For discussion, see Gregory A. Good, "John Herschel's Geology: The Cape of Good Hope in the 1830s," in J. Buchwald and L. Stewart (eds.), *The Romance of Science: Essays in Honour of Trevor H. Levere* (Cham: Springer, 2017), 135–50.

37 As David Brewster put it in his *Treatise on Optics* (1837, new edition), "Many persons have acquired the art of using this instrument with great facility, while others have entirely failed." (278)

38 Fiorentini, "Optical Instruments," 219.

39 Schaaf, "John Herschel," 88.

40 Schaaf, "John Herschel," 90. Also see Larry J. Schaaf, "The Poetry of Light: Herschel, Art, and Photography," in D. G. King-Hele (ed.), *John Herschel 1792–1871: A Bicentennial Commemoration* (London: The Royal Society, 1992): 77–100, 78.

41 Brian Warner, "The Herschel Condition," *African Yearbook of Rhetoric*, 2 (2011): 29–40, 31.

42 See Lorraine Daston and Peter Galison, *Objectivity* (New York: Zone Books, 2007).

43 Erna Fiorentini, "Practices of Refined Observation: The Conciliation of Experience and Judgement in John Herschel's Discourse and in His Drawings," in Erna Fiorentini (ed.), *Observing Nature-Representing Experience: The Osmotic Dynamics of Romanticism 1800–1850* (Berlin: Dietrich Reimer Verlag, 2007), 19–42, 28.

44 John Herschel, *A Preliminary Discourse on the Study of Natural Philosophy* (London, 1830), 3.

45 John Herschel, "Humboldt's *Kosmos*," *The Edinburgh Review*, 87 (1848): 170–229, 171–72 (italics in original).

46 Quoted in Erna Fiorentini, "Subjective Objective: The Camera Lucida and Protomodern Observers," *Bildwelten des Wissens: Instrumente des Sehens* (Berlin: Akademie Verlag, 2004): 58–66, 61.

47 Herschel, *Discourse*, 77.

48 Herschel, *Discourse*, 118.

49 Herschel, *Discourse*, 120, 118, 132. Even though in previous centuries drawing instruments were referred to as "mathematical instruments"

because they were primarily tools of land-surveying and measurement, Herschel catalogs the camera lucida a "philosophical instrument." To be philosophical meant it had reached the loftier status of a *scientific* instrument, a locution just beginning to take hold during this period. See Deborah J. Warner, "What Is a Scientific Instrument, When Did It Become One, and Why?," *The British Journal for the History of Science* 23 (1990): 83–93; and Jim Bennett, "Early Modern Mathematical Instruments," *Isis*, 102 (2011): 697–705.

50 Fiorentini, "Osmotic," 26, 28.

51 Fiorentini, "Optical Instruments," 218.

52 Herschel, *Discourse*, 120, 77.

53 Contrast this point with Marie-Noëlle Bourguet, "A Portable World: The Notebooks of European Travellers," *Intellectual History Review*, 20 (2010): 377–400. Also see Elaine Leong, "Read. Do. Observe. Take Note!," *Centaurus*, 60 (2018): 87–103. For an example of this fusion of record and attention, see Barbara Wittmann, "Outlining Species: Drawing as a Research Technique in Contemporary Biology," *Science in Context*, 26 (2013): 363–91.

54 Herschel, *Discourse*, 130 (italics in original).

55 Quoted in Usselman, *Pure Intelligence*, 74–75.

56 James Fergusson, *Picturesque Illustrations of Ancient Architecture in Hindostan* (London, 1848), iv.

57 George Airy, "History of Nebulae and Clusters of Stars," *Monthly Notices of the Royal Astronomical Society*, 3 (1836): 167–74, 174

58 Brian Warner, "The Years at the Cape of Good Hope," *Transactions of the Royal Society of South Africa*, 49 (1994): 51–66, 58–59.

59 The "barely visible" is a term of art, developed in O. W. Nasim, *Observing by Hand: Sketching the Nebulae in the Nineteenth Century* (Chicago: University of Chicago Press, 2013). It is deployed to account for scientific objects that are not visible nor entirely invisible, as most literature on visualization in the sciences has divided phenomena, but something between the two. This is important to emphasize, because, due its barely visible status, objects like the nebulae were treated in the process of visualizing and imaging in ways that were not reducible nor similar to how the invisible or visible were treated. Barely visible objects require a separate analysis.

60 R. B. Bate, "On the Camera Lucida," *A Journal of Natural Philosophy, Chemistry, and the Arts*, 24 (1809): 146–50, 149–50.

61 Cornelius Varley, *A Treatise on Optical Drawing Instruments*, (London, 1845), 33–54.

62 Years later, in 1868, French architect M. Revoili invented the *Téléiconograph*, essentially a camera lucida that could be attached to the theodolite and used for land-surveyors and military purposes. Revoili's invention, it was noted, "insures certitude in drawing, but it does not draw.

It is an aid to the artist, not a self-acting substitute for his eye and hand." See "Teleiconograph," *The Builder*, 27 (1869): 560.

63 Charles Babbage, "On the Application of a Camera Lucida to a Telescope," *Monthly Notices of the Royal Astronomical Society*, 3 (1836): 190.

64 John Herschel, *Results of Astronomical Observations Made during the Years 1834, 5, 6, 7, 8, at the Cape of Good Hope* ... (London: Smith, Elder and Co., Cornhill, 1847); hereon *Cape Results* for short. For its publication history, see Steven Ruskin, *John Herschel's Cape Voyage: Private Science, Public Imagination and the Ambitions of Empire* (New York: Routledge, 2004).

65 For more details, see Nasim, *Observing by Hand*.

66 Quoted in Augustus De Morgan, *An Explanation of the Gnomonic Projection* ... (London: Baldwin and Cradock, 1836), 104–5. Herschel was also at the forefront of the nascent astrophysics of stars, see Stephen Case, *Making Stars Physical: The Astronomy of Sir John Herschel* (Pittsburgh: University of Pittsburgh Press, 2018).

67 These "monographs" are held at the Royal Astronomical Society's archive and library. Also, the "working skeletons" referred to here should be included under the more general heading of what I have referred to as "working images" in Nasim, *Observing by Hand*.

68 Herschel, *Cape Results*, 11.

69 Herschel, *Cape Results*, 12.

70 Herschel, *Cape Results*, 40.

71 Herschel, *Cape Results*, 27.

72 Herschel, *Cape Results*, 12–13, 37.

73 See O. W. Nasim, "Astrofotographie und John Herschels 'Skelette,'" *Zeigen und/order Beweisen?*, H. Wolf, ed. (Berlin: De Gruyter, 2016), 157–78; and Nasim, "The 'Landmark' and 'Groundwork' of Stars: John Herschel, Photography, and the Drawing of Nebulae," *Studies in the History and Philosophy of Science*, 42 (2011): 67–84.

74 Letter from Sir John Herschel to Mr. Thomas Bell [1853] in Correspondence concerning the Great Melbourne Telescope, In Three Parts: 1852–1870 (London: Taylor and Francis, 1871), 22; italics in original.

75 I deal with some of these technical limitations and the ways in which they were overcome in O. W. Nasim, "Hybrid Photography in the History of Science: The Case of Astronomical Practice," in Sara Hillnhütter, et al. (eds.), *Hybrid Photography* (London: Routledge, 2021), 11–27.

76 Herschel, *Cape Results*, 29.

77 John Herschel, "Address to the Royal Astronomical Society, April 11, 1827," in *Essays from the Edinburgh and Quarterly Reviews with Addresses and Other Pieces* (London: Longman, Brown, Green, Longmans, & Roberts, 1857), 466–88, 469.

78 Contrast this to the suggested correspondence between mechanical objectivity and procedure posited in Daston and Galison, *Objectivity*, 121, 187.

79 Lisa Gitelman, *Paper Knowledge: Toward a Media History of Documents* (Durham: Duke University Press, 2014).

80 Ursula Klein, *Experiments, Models, Paper Tools: Cultures of Organic Chemistry in the Nineteenth Century* (Stanford: Stanford University Press, 2003).

7

Photology, Photography, and Actinochemistry
The Photographic Work of John Herschel

By January 1839, when the daguerreotype was first made public, John Herschel had long been familiar with the basic chemical and optical components necessary to begin his own photographic experiments. His use of visual expression in his observations, combined with his experiments on chemistry, light, and optics, had begun two decades or more before Louis Jacques Mandé Daguerre (1787–1851) and William Henry Fox Talbot (1800–77) made their respective claims to photographic invention. Because Herschel's photography has been studied largely through the lens of the history of photography, discussions are often inflected by various origin stories that privilege, for better or worse, the creation of recognizable, fixed images. This frame truncates the full and lifelong series of experiments that were eventually to become Herschel's "photography," which became a collective heading under which he consolidated his chemico-optical experiments, gathering them into a single strand of research that began as early as the 1810s and persisted through the end of the 1850s. In this extensive group of experiments, Herschel made systematic use of photography for spectral analysis and chemical sensitometric testing long before the photo-chemical work of Henry Roscoe (1833–1915) and Robert Bunsen (1811–99) in Manchester and even longer before Ferdinand Hurter (1844–98) and Vero Charles Driffield (1848–1915), who between them established important benchmarks in the nascent field of photochemistry and sensitometry.[1] Signaling that he recognized the direction his experiments on light were taking away from trends in photography by the 1850s, Herschel renamed his use of photography for light experiments "actino-chemistry," differentiating it from making optical images.[2]

The extent of Herschel's photography is perhaps less known and less understood than many of his other strands of research, making a brief tally of his output worthwhile. In total, he published at least nine important articles on photographic chemistry between 1819 and 1858. He introduced hyposulphite as a fixer and seven new imaging processes, among them the Cyanotype or, as it became commonly known, the blueprint. He produced negatives on glass, anticipating the breakout innovation of the 1850s by a decade. He is well known for popularizing important vocabulary like "photography," "snapshot," "negative," and "positive," and he was instrumental in supporting a thriving network of individuals now considered photographic pioneers. Part of the reason Herschel's work is less known is that his notebooks containing the bulk of these experiments are fragmented and largely loose-leaf rather than organized in bound volumes.[3] The photographic material likewise is scattered and is frequently misunderstood because its meaning is rooted not in image-making but in the chemical findings exhibited by the spectra, negatives, and prints.[4] Mike Ware made an initial attempt in 2006 to determine the volume of Herschel's photographic output, counting 771 paper preparations created up to the year 1843. He also noted that each paper could be divided into twelve to sixteen pieces.[5] Even using the conservative estimate of twelve, that puts the number of Herschel's photographs at comfortably over nine thousand in just five years, and he continued to produce them off and on for more than a decade. To understand this enormous bulk of Herschel's photographic work, it is important to foreground chemical experiment rather than image-making, even though Herschel was well aware of the power of images as analytical tools. It makes sense then to first investigate the roots of his interests, as they remained remarkably consistent for nearly thirty-five years.

Photology

As early as 1813, Herschel was deeply invested in what he titled "optical, chemical & nonsensical" investigations, pursuing many of them "with [Charles] Babbage" or at Babbage's laboratory. These experiments cover a wide range of topics, from detonating phosphorous mixed with hyperoxymuriate of potash to attempting to make sulphuret of carbon.[6] Many of the substances Herschel used at the time are now familiar to photographic historians: muriate of silver, phosphorous, sulphuric acid, oxalic acid, muriate of lead, and more. In 1819, he began publishing a series of articles on hyposulphurous acids.[7] This research formed the basis of his knowledge in the use of hyposulphites, and in 1819 Herschel tested the dissolving and precipitating powers of his hyposulphite on lead, silver, mercury, and a number of other substances. In 1820, Herschel published again, this time on the refractive power of hyposulphite of soda and silver nitrate of lead.[8] His knowledge of and interest in the refractive power of chemical combinations underpinned later photographic experiments on vegetable colors (see Appendix at end of chapter) and various metals.

By the middle of the 1820s, Herschel's experiments in precipitation of different metallic and hyposulphurous compounds dominated his chemical investigations. A typical experiment from 1825 goes like this:

Oct. 8. 1825
952. a) Nitrate of Lead being precipitated by Ferrocyanate of Potash, a Ferro-cyanate of Lead was prepared which was a white, very finely divided powder. b) This was treated with Sulphurated Hydrogen which decomposed it and the Sulphuret of lead being sep[d] by filtration – an Acid liquor was procured which from the mode of its formation ought to be solution of Ferro-cyanic Acid. Its taste purely acid not disagreeable c) This sol[n] exposed to the air underwent a slight decomp[n]. & acquired a shade of blue. Heated on a sand bath, it decomposed rapidly & a good deal of Prussian blue separated.[9]

He further noted that this combination "decomposed" all his metallic salts, giving definite colors: blue for persalts of iron; brown for copper and uranium; white for manganese, mercury, cesium, and lead; and green for nickel. Herschel's use of the persalts of iron, and the deep blue color it turned, would be characteristic of his Cyanotype process not quite twenty years later. Likewise, this preoccupation with the color resulting from chemical combinations demonstrated his belief that color could be deployed as an analytic tool of some sort, even if it could not be relied on as an inherent quality of an object. Even as he established the inconsistency of color, he continued to investigate its ever-changing nature by carefully noting color changes in his experiments.

The research on hyposulphurous acid and its compounds and Herschel's interest in color change laid a foundation for further chemical investigation by light, heat, and chemical intervention. In 1831, he added to this knowledge a substance new to him, found amongst his father's chemicals, which he called "platinate of lime" (calcium chloroplatinate) and which could be darkened in the presence of light.[10] Herschel demonstrated this experiment to Babbage, Talbot, David Brewster (1781–1868), and several others who had gathered at Babbage's house for breakfast on June 21, 1831.[11] Leading up to this demonstration, throughout March and April 1831, Herschel also explored the effect of colored filters on the experiment. He immersed the bottle of the light-sensitive substance in a larger jar filled with liquids that he had made by dissolving various materials, including one of a tincture of red rose leaves, to create a red filter. Herschel's investigations of the properties of vegetable colors first as filters and later as objects of inquiry were to grow in volume and importance through the decades. In the series of platina experiments, Herschel began assigning his substances new names, "photo-platinate of lime" and in the following experiment "photo-platinate of silver," indicating that he was beginning to think of such experiments and substances as a class of their own.[12] By 1832, after considerable advancement in his understanding of the action of light on chemical substances, Herschel applied a general title, "photology," to describe the entire set of experiments.[13] It was a word applied now and again to studies on light, and Herschel even directed this term at Talbot's nascent experiments in a letter of 1837.[14]

Although Herschel did not invent the term "photography," he used it in his first public presentation of his photographic work in March 1839 at the Royal Society. Others had suggested it as a name before, but it is clear that Herschel's use of it at the Royal Society popularized the term, even if it was unevenly adopted in practice.[15] This was part of Herschel's attempt to form a new area of study through systematic experimentation in all its areas, complete with a practical and meaningful scientific nomenclature. Naming was a critical aspect of organizing knowledge during this period that Herschel shared with contemporaries like Brewster and William Whewell (1794–1866). That Herschel considered the chemico-optical experiments as a class of experimentation all their own is an important indication of how far his experiments of 1819 and 1820 had led him and how critical an understanding of this decade is for understanding Herschel's photography after the announcement of the daguerreotype.

Photography

When François Arago (1786–1853) announced the daguerreotype to the *Academie des Science* on January 7, 1839, it was sensationally (and often inaccurately) reported in the press on both sides of the English Channel. No actual chemical details would emerge until August of that year, so eager experimenters were left to employ their own imaginations and draw on their own chemical knowledge. News traveled rapidly in letters between scientific families, and it was most likely the personal communication of Sir Francis Beaufort (1774–1857) that alerted Herschel to the daguerreotype later in the same month.[16] His notebook details how Herschel was able to unravel "Daguerre's secret," as he called it, in a matter of days around January 29. It took only two experiments for Herschel to note that "Hyposulphite of soda arrests the action of light," fulfilling one of the three prerequisites he had outlined for successful imitation of Daguerre's process. Although the other two prerequisites, "very susceptible paper" and a "very perfect camera," were no doubt critical as well, it was this fixing of the image that emerged as the material form of "success."[17] It is a fault line that has also directed the writing of the history of photography ever since, with experiments before fixing sometimes given the title "proto-history" and those after it designated the "history" of photography.[18] Fixation of the image was one of only two characteristics that Talbot used in letters to Arago and Alexander von Humboldt (1769–1859) to establish his own invention priority.[19] Fixer, or rather the hyposulphite solution, also made a physical impression on the notebook pages where Herschel's photographic experiments begin, in the form of a large purple stain visible in the lower half of the gutter of the notebook for four pages between January 29 and February 13, accompanied by its distinctive smell. It is thus surprising that Herschel's experiments on the development of hypo are little discussed. Though this chapter does not have the space to address the full scope of these trials, they are important for establishing two things: the collaborative nature of photochemical investigation and the important role that color played in evaluating "successful" fixing.

Both notes and images survive from Herschel's fixing experiments that immediately call into question their designation as "Herschel's photography," because Herschel experimented not only on his own papers but also on images and papers sent to him by others, mainly Talbot and the chemist and science writer Robert Hunt (1807–87). Larry Schaaf has written extensively on the collaborative nature of Talbot's and Herschel's experiments, and it is important to understand that this sort of collaboration was not unusual for Herschel.[20] Hunt too was interested in the chemical action of light on the spectrum and presented his own research and direct positive process in 1840, though these are often eclipsed by his more influential books documenting the history of photographic experiment.[21] Hunt exchanged many letters and specimens with Herschel, discussing observations on light, achievements by various chemists and experimenters, and advances in research.

Herschel's *Photographic Memoranda* notebook contains not only a complete list of his photographic paper preparations, it also demonstrates, in experiments numbered 710 through 714, his trials "fixing" Hunt's "blank photogenic paper" in several different ways (Figure 7.1). He proceeded by testing not only the hyposulphites but by trying to replicate other methods, for instance, the use of hydriodate of potash and chromate of potash as well as washing with water. The experiments are not only directed at gaining a photograph capable of withstanding further exposure to sunlight; achieving the right color was important too. Talbot, Herschel, and others had immediately noticed that the delicate colors of photogenic drawings were obliterated by fixing in hyposulphite. Unfixed or stabilized prints range in color from a lilac to an orangy-brown. But not all substances had the same effect. Chromate of potash, Herschel noted, did fix photographs but rendered the paper a "nasty thick orange,–dirty looking colour." Attempts underlined by Herschel, clearly indicating more success, rendered photographic results that looked black, very much like a "real print."[22]

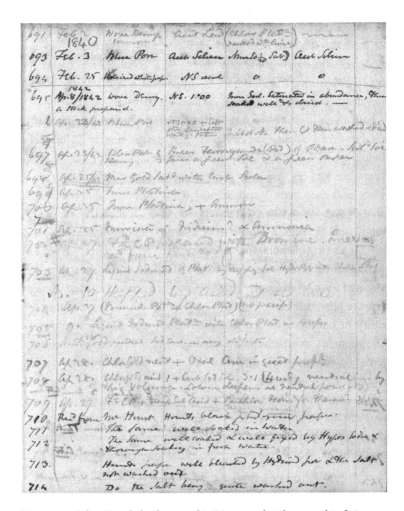

Figure 7.1 John Herschel. *Photographic Memoranda*, photographic fixing experiments 710–714. c.1839 (Harry Ransom Humanities Research Center, Austin, Texas)

Attaining photographic colors that imitated prints was one clear trajectory of Herschel's experiments. The other was to make copies from engravings efficiently. In what was only his second recorded day of photographic experiments, January 30, 1839, Herschel had already begun to use engravings on paper as negatives, often oiled to aid transparency. These semi-transparent engravings were then placed directly in contact with the photographic paper, squeezed under a sheet of glass, and placed in the sun for exposure. When they had exposed long enough, they were fixed and washed variously depending on the experiment.[23] The chemistry of silver and other metal salts reacts to light by turning dark. Thus, the prints Herschel made were, in his own words, "reversed" in comparison to the original print. On the very next day, after he had made his first successful "reversal" or "copy" of an engraving, he "re-reversed" the copy of the print, rendering it back in its original tones. The reversal and re-reversal of prints occupied not just his photographic experiments but his attempts to form a vocabulary for these new photographic activities.

Though his addition of nomenclature to photography may seem trivial compared to fixer or the Cyanotype, it is worth considering how his naming conventions reflect his important role in the midst of a burgeoning network of photographers. At first Herschel's vocabulary consisted of "transfers" and "re-transfers," "reversal" and "re-reversal," sometimes of "colours" and sometimes of "tints." This convoluted way of discussing darks and lights could lead to quite lengthy descriptions and permutations like double transfers, transitional photographs, and original transfers. Eventually, in 1840, Herschel borrowed electrical terminology and applied "negative" and "positive" to describe photochemical activity and give it a regularized vocabulary.[24] Again, he chose the Royal Society to present his new findings and vocabulary, and as his first published paper on photography it carried weight. Herschel's important position in the network of photochemical investigators allowed him insight into the breadth and depth of the field and gave him a reason to facilitate exchanges by simplifying the way photography and its methods were discussed. Regularizing nomenclature and giving photography a vocabulary of its own set it apart as a field and shaped those practicing it into a community who shared tacit practices and a set vocabulary.

Both the 1839 paper "Note on the Art of Photography" and the 1840 paper "On the Chemical Action of the Rays of the Solar Spectrum" demonstrated the power of linking vocabulary to practice even if the first paper was never published.[25] In the first paper, Herschel offered "photography" as his name for the research area and provided two demonstrations of its capabilities. The first was affixing a spectrum to the manuscript as evidence of the difference in exposure under glass.[26] The paper also served to introduce members of the committee of the Royal Society to this new thing called "photography," as not all the members may have seen a photograph. The second was providing an album of photographic specimens accompanying the paper. I call it an album, perhaps the first of its sort, because of its careful preparation and captioning and its being bound up in a book-sized object. This small notebook, titled "Specimens of Photography," has a marbled cover, and each photographic specimen is carefully affixed to a single page, centered and cut with a border, and each bearing a handwritten caption beneath. Most of the images have faded entirely, leaving precisely trimmed rectangles of white poignantly affixed to the white notebook pages above carefully written captions.[27] At the time these multiple uses of photography, affixed to the manuscript and accompanying it in an album, established it as a new sort of reliable demonstration of scientific experiment. Perhaps in recognition of the importance of these photographic offerings, the first image Herschel attached in the album is one he made, employing a self-constructed camera, of the structure of his father's famous forty-foot telescope.[28]

Telescope Images

The telescope may seem an odd choice as a first image, especially as the majority of Herschel's photographic output was spectra, though he also made a number of positives and negatives of lithographs and engravings and produced smaller numbers of images made by placing leaves, feathers, and other objects directly on paper or by using a camera. But as soon as he began experimenting with photography, Herschel started a series of photographs of the telescope scaffolding that would continue through the summer and autumn of 1839, until the huge relic was dismantled and the Herschel family moved from Observatory House in Slough to Collingwood in Kent. On January 30, his second recorded day of photographic experimentation, Herschel set to work constructing his first camera and making what he termed an "optical image" of the immense forty-foot telescope:

> Jan. 30/39 Formed images of telescope with the aplanatic lens of Dollands make [on] my construction ()(and placed in focus paper with Carb Silver. An image was formed in <u>white</u> on a sepia-col^d ground after

about 2 hours exposure which bore washing with hypos. Sod. & was then no longer alterable by light. – Mr Daguerre's probl^m is so far solved – for granting a perfect picture, the picture of this may be taken.[29]

Herschel, like many early photographers who constructed their own cameras, used an aplanatic lens to form a narrowly focused image on paper. With a two-hour exposure, the structure of the telescope mount was the perfect static subject. It was also easily accessible from the house, which served both as a platform to raise the camera to the right height and as a darkroom for fixing and washing the resulting image once it was removed from the camera. The frame of the immense telescope was already part of Victorian visual culture, having been engraved several times and depicted in at least one encyclopedia. No matter how soft-focused, the shape of the telescope would have been unmistakable to nineteenth-century scientific viewers. Herschel eventually took more than fifty of these negatives on paper and glass (and made numerous prints), first from a lower-level window and later from at least two different upper ones, continuing to photograph the instrument through the end of October 1839.

These near-identical, often round photographs have attracted little attention historically, although they represent an important phase of Herschel's photography. They were significant enough for him to take two of them, made on his newly discovered plumbozoic paper, as gifts to Arago and Jean-Baptiste Biot (1774–1862) when Herschel visited Paris in May 1839.[30] The presentation of these visual mementos of the telescope, which had by that time been dismantled, was not unlike later scientific exchanges of carte de visite portraits, incorporating both celebrity and status.[31] Serial photographs of a single unmoving target also allowed him to exert some control and compare his photographic experiments to one another. Photography was capricious, and its variables made it difficult to quantify experimentally. There was, for instance, no term or measurement for levels of chemical change or so called "actinic" activity. Different paper surfaces elicited changing results. The strength of the sunlight came and went with the cloudy English weather. Herschel relied instead on photographing the same subject in series, and he was to continue with this technique through the 1840s and 1850s, when he often used the same engravings again and again.

The most well-known of Herschel's telescope photographs are a series of several round glass negatives, made by precipitating muriate of silver directly on the glass for forty-eight hours, drying, washing with a weak silver nitrate solution, and drying again. It was a method Herschel devised that was so delicate only a few other scientists tried it, and even fewer succeeded. These photographs are often referred to as the "first" negatives on glass, sometimes in a way that implies he made only one.[32] The image could be rendered positive two ways. The first was by the addition of a frosted piece of backing glass, with the whole package sandwiched together with a small paper spacer to separate the glass pieces. Herschel called this a daguerreotype on glass, due to the ability to turn it into a direct positive by blacking the back.[33] Secondly, Herschel also printed some positives through at least one of the glass negatives in the years following 1839, adding them to the paper negatives of the telescope. It is difficult to know how many glass negatives Herschel made, as he notes that some were less successful and at least one broke. Perhaps the total number was no more than three or four.

Though the telescope images make up only a small portion of Herschel's photography, they are the first sign of his technical tinkering with photographic apparatus. He also invented a method for a solar reducer to form smaller images, and theoretically he could have reversed the apparatus to enlarge them too.[34] There are few notes of the camera Herschel used in 1839, aside from the lens, but it was the first of several types of camera he developed, most of them in service of his longest-running photography experiments using spectral analysis. He would eventually name his setup the "spectroscopic camera" using various iterations of it to make photographic spectral analysis over several decades.

Actinochemistry

Herschel's photographic spectra are widely dispersed and most are missing. He occasionally pasted spectra onto letters, and some are pasted into Notebook 3 at the Science Museum. This haphazard scheme of keeping his spectra results and the dispersed nature of his notes make it difficult to understand how Herschel organized his photographic outputs with his notetaking. Photography disrupted the usual pattern of many scientists, often being incompatible in a way drawings were not.[35] Sometimes photographs were made on incompatible substrates like silver or glass plates, but often it was merely the extended time required for the creation, exposure, and subsequent treatment of photographic materials (sometimes development, but always fixing and washing). Like many scientists, Herschel often drew sketches of his photographs, using a graded cross-hatching to indicate strength or weakness of the photographic exposure. There are some indications that Herschel developed a systematic ordering system for keeping photographic spectra. One remaining notebook in the Harry Ransom Center holds 189 spectra made between April and September 1859 pasted into the Royal Society's 1855 Membership List. The spectra are inserted in number order, with the number at the top, the spectrum in the middle, and notes about preparation at the bottom. Occasional numbers at the top, just under the experiment number, indicate paper preparations. Herschel also conserved even those spectra that were unsuccessfully fixed on paper (Figure 7.2). In the three cases of 1281, 1282, and 1283, the spectrum washed out of the paper, which was not an uncommon occurrence. Nonetheless, Herschel noted the length of the spectrum before it was washed away.

Herschel's spectral analyses of 1858–59 were the culmination of decades of work on hundreds of preparations that included changes of metal, changes of paper, experiments on sizing, and trials of fixing methods. Not only were there different metals to try but different salts to combine them with, innumerable plants from which to extract dyes, several sizing agents, and countless combinations of surfaces on which to put the concoction. Herschel realized very early that the make and finish of the paper, as well as the sizing agent, affected photographic activity.[36] (It was often the sizing agent that caused images to wash off paper, too.) His notebook contains a list of trial papers, including thin and thick wove post, demy, gilt-edged paper, china paper, letter paper and engraving paper. On these papers, he mixed his light-sensitive solutions with various sizing agents, including gum water, solutions of soap, starch, isinglass, and egg albumen. He also varied the substrate, trying photographs on ivory, parchment, goldbeater's skin, and porcelain biscuit.[37] The spectra experiments began as early as July 6, 1839. On July 7, he built another apparatus using a Fraunhofer crown glass prism and a glass lens to make a very concentrated spectrum, which he was able to photograph using Talbot's muriated paper. The colors, which Herschel called tints, were subtle (see Appendix at the end of the chapter) and fugitive.[38] In order to improve the representations, he

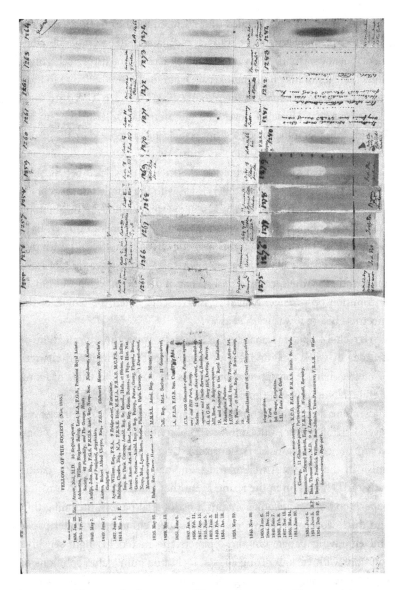

Figure 7.2 Photographed spectra in "Experiments on the Photographic
Sensibilities of Salts of Supposed New Elements Made with the Spectroscopic
Camera by Sir J. Herschel Bart. at Collingwood April–Sept. 1859." Harry
Ransom Humanities Research Center, Austin, Texas.

tinkered with the chemistry as well as the technology, employing at times a water prism and a flint glass prism in combination with lenses to "intensify" the spectrum. To stabilize the long exposures necessary, he eventually added a clock mechanism to alleviate the need to adjust the apparatus continuously by hand.

The spectra experiments are equally representative of Herschel's continued pursuit of a working theory of light and a new interest in fixing natural colors.[39] The experiments derive from his February and March 1839 experiments with the so-called exulting glass, combined with his use of projective observation of spectra and polarization.[40] The discovery of new facts about light through absorption that seemed so promising then were not forthcoming. Although Herschel devoted many days of experiment to the apparent action of colored glass on the photosensitive effect, the results were inconclusive. Herschel presented the first findings of his exulting glass experiments at the Royal Society in March of 1839, in the paper that he subsequently withdrew from publication. The exulting glass experiments, which had their roots in Herschel's 1820s and '30s research and publications on absorptive spectra, did not explain absorption spectra and thus led to no new explanations of undulatory theory.[41] Unsurprisingly, these early experiments figure less prominently in his first published paper on photography a year later in 1840.[42] In part, Herschel laid the blame for his lack of general conclusions in 1840 on the capaciousness of the spectra and the many directions of inquiry they encompassed. It would take another two years for him to consolidate his findings and make his new processes known.

Photographic Processes

In total, Herschel invented and named six or, depending on how they are counted, seven processes using different metal salts.[43] Among these were the Kelaenotype or Celaenotype (mercury), Siderotype (iron), Amphitype (mercury), Argentotype (silver-iron), Anthotype or Phytotype (plant), Crysotype (gold), and the most influential, the Cyanotype (iron). These processes are not just an indication of Herschel's ingenuity, they are evidence of his professional networks in the growing field of industrial chemistry. The history of the Cyanotype has been thoroughly explored by Ware, Schaaf, and others, and with good reason. Under its more common name, the "blueprint," it became the dominant method of industrial and architectural copying for nearly a century. The details of its invention are deeply entwined with parallel innovations made through experiments on electricity and the growing industrial use for synthesized chemicals.[44] In 1839, photography arrived on the heels of almost forty years of electrical research in Britain and elsewhere. Electricity had become particularly important for isolating purified chemical compounds, as Humphry Davy had already proven. There was also a noticeable increase in the number of industrial chemists in Britain, accelerated by the 1845 founding of the Royal College of Chemistry in London.

Alfred Smee (1818–77), a surgeon and chemist working for the Bank of England, is an example of this sort of industrial chemist whose knowledge and chemical activities directly supplied materials for Herschel's photographic experiments. Smee was Herschel's supplier for refined chemical substances, like the potassium ferricyanide that led to the Cyanotype. Smee created it by applying electricity from a battery of his own invention to preparations of iron, synthesizing potassium ferricyanide in a very pure form from potassium ferrocyanide. Soluble potassium ferricyanide appeared under different names in Herschel's notebooks and publications; red ferrocyanate of potash or ferrosesquicyanuret of potassium are used equally. Smee provided Herschel with a sample in the spring of 1842, and Herschel used it to produce many papers with a "strong blue impression," leading to a veritable cornucopia of photographic processes.[45] Among them were more than a dozen viable chemical formulae for the Cyanotype, which Herschel explained as a class of processes rather than a single process, in which cyanogen and iron are combined and the resulting photograph is Prussian blue.

Both the Cyanotype and the Chrysotype, another process that used additional gold salts, were presented toward the end of Herschel's 1842 paper and given more detail in a lengthy appendix. The paper was accompanied by photographs to illustrate his findings, presented four to a page with instructions for viewing. In the case of the Crysotypes (Figure 7.3), Herschel exhorted viewers to use good light and magnification to observe the advantages of this beautiful process. The majority of the 1842 paper concentrated on plant material (see

Figure 7.3 John Herschel, Chrysotype images supplied with Art. 212, Herschel. 1842 paper (Royal Society Library)

Appendix) and their colors, as well as their permanence or impermanence. Photographs with plant substances and those using metal salts were not put together by accident. The organic ferric salts were, at the time, referred to as vegetable acids and had been in use at least since the middle of the eighteenth century.[46] The Chrysotype and Cyanotype, a deep purple and deep blue respectively, also fit into the category of highly colored photographs; unlike the Anthotypes, however, they were remarkably permanent. The plant substances Herschel used and the Prussian blue he created with Smee's ferricyanide were both active ingredients in the dyeing industry and had been the target of industrial synthesis for decades. That these substances were in common circulation and the subject of chemists' experiments explains Herschel's choice to pursue them. Herschel's process inventions demonstrate the centrality of industrial chemistry to photographic innovation.

However, the Cyanotype should not be seen as only an object of Herschel's experiments. It also became a locus for sharing photographic knowledge and useful for organising observations. The Cyanotype could be used to make a series of "blanks," as, for instance, forms for recording sunspot observations between December 1869 and May 1871, just days before Herschel's death (Figure 7.4). The Cyanotype was not used to photographically record the sunspots themselves but to create a form with directional arrows and other indications and a blank circular interior for recording visual observation in ink. The bottom half of each paper still bears the characteristic blueish stains of Cyanotype chemistry, not fully washed out. By this time, Herschel would not have needed to make his own cyanotypes, as his daughter Julia (1842–1933) and other members of the family were proficient with photographic chemistry. Julia, in fact, authored a book in 1869 about Greek and Roman lace using Cyanotype illustrations.[47] In sharing his photography with his children, Herschel was carrying on his tradition of demonstrating photographic action at work.

The collaborative nature of Herschel's photography extended in many directions and included both men and women in sharing technical tips, giving and receiving photographs, supporting the work of photographic experimenters, and collecting photographs and important publications about them. It is difficult to quantify the size and shape of Herschel's influence, but evidence can be found in his activities of 1839, in his promotion of photography innovators outside the family as well as within it – among them Robert Hunt, Anna Atkins (1799–1871), and Julia Margaret Cameron (1815–97) – and in the Herschel collection, which although dispersed has informed the writing of photographic history. Herschel was aware of most of the experimenters working on early photography, both the daguerreotype and paper processes, and was quick to acknowledge that his own

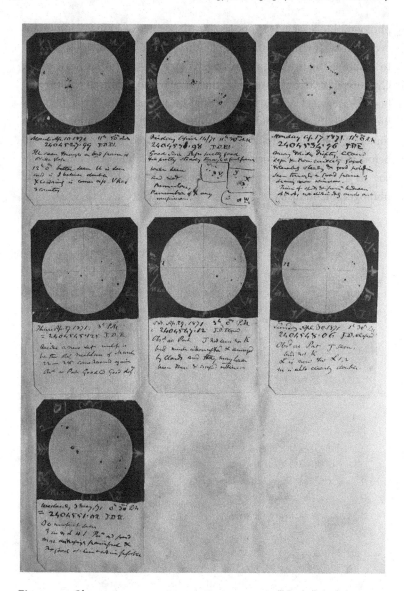

Figure 7.4 Observations on sunspots using cyanotype "blanks" or forms (Royal Astronomical Society. Observations made in May 1871)

photographic knowledge depended on the prior experiments of men and women of science like Elizabeth Fulhame (fl. 1794), whose work he acknowledged in his March 1839 paper. It is, however, Herschel's intangible influence that constitutes the most intriguing part of his photographic legacy. He appears near the center of so many important events in photographic history, like the French government's decision to purchase the invention of photography and Anna Atkins' use of the Cyanotype to create arguably the first photographically illustrated book.[48] Herschel remained influential in photography throughout his life, contributing new findings and consolidating long-running research topics in spectral analysis. Although his research was conducted well before the field of photochemistry emerged, it can be seen as a precursor to a field that was firmly embedded between science and the photographic industry.

Appendix

Herschel's Vegetable Colors

In the two years between 1840 and 1842, Herschel experimented extensively with plant materials. Using "vegetable colours," he studied photographic color changes exhibited by tints extracted from resins, flower petals, and leaves from trees and vegetables.[49] From this investigation, he developed the Anthotype process, which employed the fading of plant tints to produce a positive photographic impression. Impermanent and with lengthy exposure times, this method proved to be an impractical road to color photography. Spectral analysis formed the core of the investigation, demonstrating how deeply Herschel's photographic experiments were embedded in his research on light and how plants were employed to render properties of light visible (see Figure 7.5).

In his first publication on photographic experiments in the 1840 *Philosophical Transactions*, Herschel stated his motivation to study the well-known yet "un-investigated" fading process in "vegetable colours." His discovery of the "Herschel Effect," in which the red region cancels out the photographic effect of the blue region on silver-chloride paper, raised the question of what kind of changes plant tints undergo if exposed to the spectrum and whether they "moreover, by such changes indicate chemical properties in the rays themselves hitherto unknown."[50]

The two-year-long investigations that followed formed the main part of his second article in the 1842 *Philosophical Transactions*. The article presented a selection from his over 100 experiments and 225 paper preparations with an extensive spectral analysis of plant tints.[51] Spectral analysis, in which a projection of the solar spectrum

Figure 7.5 Reworking Herschel's experiment 1183 with elder leaves using a modern reconstruction of Herschel's "spectroscopic camera" and plant-based chemistry (Carolin Lange, August 6, 2022, 10:30–11:30 a.m.)

is created with a combination of prism and lens, was well-established in optical physics of the eighteenth and nineteenth centuries, with many scientists using photosensitive substances to investigate light and matter.[52] Highly photosensitive silver salts produced photographic impressions on paper within seconds to minutes. "Vegetable colours," however, needed exposure times of up to several hours to create a visible photographic effect. During long exposures, Herschel observed how the solar spectrum created and destroyed colors in a performance of subtle photographic impressions, positive and negative, changing in color and intensity and often of a complementary nature. It was a performance that attributed chemical activity to all spectral regions but could not be fixed photographically.

Herschel's selection of plant material was not unusual and connects to optical research traditions in a way that indicates his hopes for eliciting more general observations about light and color from this work. His experiments with the green juice from elder leaves, for instance, reson-ates with the methods and materials from Brewster's 1834 work on absorption spectra, which was itself a critical response to Newton's description of "the green of vegetation."[53] Herschel's experiments with the resin of the Guaiacum tree built on the 1802 observations of William Hyde Wollaston (1766–1828), confirming its reaction to infrared radiation.[54] Edmond Becquerel (1820–91) repeated and extended Wollaston's and Herschel's experiments with more precision in 1842. Like Herschel, Becquerel detected chemical activity in all regions of the solar spectrum using metal substances and Guaiacum resin, but only Becquerel was able to capture this activity by creating the first photographs of the solar spectrum and its characteristic dark lines.[55]

Herschel halted his experiments on plant material in 1842. A year later, he returned the financial support from the Royal Society granted to pursue his experiments due to Becquerel's "definite and important, and so far final" results.[56] Becquerel's conclusions indicated a new direction of study and not the general rules at which Herschel was aiming. The vegetable colors, however, remained stubbornly particular. In the end, Herschel's experiments with plant materials were inconclusive, as were many of his chemical experiments, though he was working in the direction that science and color photography would eventually take – namely, spectroscopy and the application of complementary color reactions.

Notes

1 Robert Bunsen and Henry Roscoe, "Photochemical Researches," *Quarterly Journal of the Chemical Society of London*, 8 (1856): 193–211; and Vero C. Driffield "The Hurter and Driffield System: Being a Brief Account of Their Photo-Chemical Investigations and Method of Speed Determination" *The Photo-Miniature*, 56 (1903): 300–41. See also Klaus Hentschel's argument that Herschel's prismatic analysis predates spectroscopy in *Mapping the Spectrum: Techniques of Visual Representation in Research and Teaching* (Oxford: Oxford University Press, 2002).

2 Stephen Case, *Making Stars Physical: The Astronomy of Sir John Herschel* (Pittsburgh: University of Pittsburgh Press, 2018), 185–86.

3 Mike Ware, "Herschel's Chrysotype: A Golden Legend Retold," *History of Photography*, 30 (2006): 1–24, on 3–4.

4 The largest collections of Herschel's photography can be found in the Harry Ransom Humanities Research Center (HRC) in Austin, Texas; at the Royal Society in London; at the Museum of the History of Science in Oxford; and

in the Science Museum Group, at the Science Museum, London; and the National Science and Media Museum, Bradford.

5 Ware, "Herschel's Chrysotype," 4.

6 John Herschel, Notebook 1, March 7, 1814. Experiments 34 and 35, Science Museum Group Collection, MS/0478/1. https://collection .sciencemuseumgroup.org.uk/documents/aa110075238/notebook-volume-1 (accessed February 1, 2023).

7 John Herschel, "On the Hyposulphurous Acid and Its Compounds," "Additional Facts Relative to the Hyposulphurous Acid," and "Some Additional Facts relating to the Habitudes of the Huposulphurous Acid, and Its Union with Metallic Oxides," *Edinburgh Philosophical Journal*, 1 (1819): 8–29 and 396–400 and 2 (1820): 154–56.

8 John Herschel, "The Refractive Power of Hyposulphite of Soda and Silver Nitrate of Lead" *Edinburgh Philosophical Journal* 2 (1820): 114–21.

9 John Herschel. Notebook 3, p. 332. Science Museum Group Collection, MS/ 0478/3. https://collection.sciencemuseumgroup.org.uk/documents/ aa110075242/notebookvolume-3 (accessed 2 February 2023).

10 John Herschel, "On the Action of Light in Determining the Precipitation of Muriate of Platinum by Lime-water," *Philosophical Magazine*, 1 (1832): 58–60.

11 Larry Schaaf, *Out of the Shadows: Herschel, Talbot and the Invention of Photography* (New Haven: Yale University Press, 1992), 33.

12 Herschel, Notebook 3, the second page of exp. 999 continued and exp. 1000 extending over pages 352 and 353.

13 Herschel, "On the Action of Light."

14 Herschel to Talbot, January 29, 1837, *The Correspondence of William Henry Fox Talbot*, doc. no. 03442, www.foxtalbot.dmu.ac.uk (accessed February 2, 2023).

15 Both the luckless Hércules Florence (1804–79) and Charles Wheatstone (1802–75) used this title before or independently of Herschel. Herschel's March 1839 paper was read before the Royal Society but not published. See Larry Schaaf. "Sir John Herschel's 1839 Royal Society Paper on Photography," *History of Photography*, 3 (1979): 47–60.

16 Schaaf, *Out of the Shadows*, 48.

17 Herschel, Notebook 3, January 29, 1839, 1013.

18 Kelley Wilder, "Invention of Photography," in Robin Lenman (ed.), *The Oxford Companion to the Photograph* (Oxford: Oxford University Press, 2005), 314–17.

19 Both letters are dated January 29, 1839, and substantially duplicate one another. Talbot to Arago, doc. no. 03777, and Talbot to Alexander von Humboldt doc. no. 03778 *The Correspondence of William Henry Fox Talbot*, www.foxtalbot.dmu.ac.uk (accessed February 2, 2023).

20 Schaaf, *Out of the Shadows*, 50 and fig. 27.

21 James R. Ryan, "Placing Early Photography: The Work of Robert Hunt in Mid-Nineteenth-Century Britain," *History of Photography*, 41 (2017): 343–61.

22 Herschel, Notebook 3, February 14 and 16, 1839, exp. 1023 and 1026.

23 In the twentieth century, these became commonly known as contact prints. Herschel was not using developer at the time, as the chemistry was printing-out not the later developing-out chemistry.

24 John Herschel, "On the Chemical Action of the Rays of the Solar Spectrum on Preparations of Silver and Other Substances, Both Metallic and Non-Metallic, and on Some Photographic Processes," *Philosophical Transactions of the Royal Society of London*, 130 (1840): 1–59.

25 Schaaf, "Herschel's 1839 Royal Society Paper on Photography."

26 For a discussion of experimental photographic evidence, see Kelley Wilder and Martin Kemp, "Proof Positive in Sir John Herschel's Concept of Photography," *History of Photography*, 26 (2002): 358–66.

27 The front cover of this notebook has been reproduced in Nadja Lenz, "The Hidden Image: Latency in Photography and Cryptography," *PhotoResearcher*, 17 (2012): 11, fig. 6.

28 It has now faded but was reproduced when visible by Schaaf in *Out of the Shadows*, 56, fig. 30.

29 Herschel, Notebook 3, January 30, 1839, exp. 1013.

30 Herschel, Notebook 3, June 4, 1839, exp. 1043.

31 Geoff Belknap and Sophie Defrance, "Photographs as Scientific and Social Objects in the Correspondence of Charles Darwin," in Elizabeth Edwards and Christopher Morton (eds.), *Photographs, Museums, Collections: Between Art and Information* (London: Bloomsbury, 2015): 139–56.

32 Not only did Herschel make several of these, but they are not the first attempts to use glass as a substrate. Nicéphore Niépce made several successful glass negatives by a different process before his death in 1833.

33 Herschel, "New Daguerreotype on Glass," Notebook 3, September 9, 1839, p. 397. Indeed, after the 1851 announcement of collodion on glass images, some daguerreotype studios began selling Ambrotypes, which were sometimes marketed as daguerreotypes on glass.

34 Herschel, March 25, 1839, Notebook 3, 1031.

35 See Kelley Wilder, "Photographs, Science and the Expanded Notebook," *Umeni Art*, 3 (2022): 279–89.

36 Sizing agents like vegetable starches or animal byproducts (gelatin or isinglass) not only made paper smoother, they also helped control the absorption qualities of ink, paint, or in this case, photographic chemicals.

37 Herschel, "Photographic Memoranda."

38 Herschel, July 6, 1839, Notebook 3.

39 Hentschel characterizes the spectra photographs as a separate activity only about fixing color in *Mapping the Spectrum*, 203ff, but Herschel's notes to his experiments appear to refer to both motivations.

40 Kelley Wilder, "A Note on the Science of Photography: Reconsidering the Invention Story," in Tanya Sheehan and Andrés Mario Zervigón (eds.), *Photography and Its Origins* (New York: Routledge, 2015): 208–21.

41 Herschel's observations of the 1820s were published in "Light," in *Treatises on Physical Astronomy, Light and Sound Contributed to the Encyclopedia Metropolitana* (Glasgow: R. Griffin and Co., 1829–43). For a discussion of Herschel's absorption work, see Frank James, "The Debate on the Nature of the Absorption of Light 1830–1835: A Core-Set Analysis," *History of Science*, 21 (1983): 335–67.

42 Herschel, "On the Chemical Action of the Rays of the Solar Spectrum".

43 Herschel commonly used more than one name for a similar process, leading Ware to count a total of nine processes, but I have made a conservative count here. It could easily be more.

44 Ware emphasizes the exciting new field of electrochemistry, which was a part of this larger industry of synthesized chemistry and photography, the production of aniline dyes being another. Mike Ware, *Cyanomicon: History, Science and Art of Cyanotype: Photographic Printing in Prussian Blue* (Buxton: Mike Ware, 2014), on 27–28.

45 Ware, *Cyanomicon*, 28. Schaaf, *Out of the Shadows*, 127–30.

46 Ware, "Chrystotype," 2.

47 Julia Herschel, *A Handbook for Greek and Roman Lace Making* (London: R Barrett and Sons, 1869).

48 Communication of F. Arago to the Académie des Sciences, using Herschel's effusive praise of the daguerreotype to argue the case. As reproduced in *Niépce Correspondence et Papiers*, document number 623, 1165. Anna Atkins self-published *British Algae* beginning in 1843.

49 For an overview of all plant materials, see Herschel, "Photographic Memoranda."

50 Herschel, "On the Chemical Action of the Rays of the Solar Spectrum," 34. For the "Herschel Effect" see Hentschel, *Mapping the Spectrum*, 204.

51 See John Herschel, Notebook 3, "Photography. Experiments and observations vegetable colours," W0278. Container 9.32. HRC.

52 See Alan E. Shapiro, *Fits, Passions, and Paroxysms: Physics, Method, and Chemistry and Newton's Theories of Colored Bodies and Fits of Easy Reflection* (Cambridge: Cambridge University Press, 2009); and Case, *Making Stars Physical*, 167–98. These scientists included Carl Wilhelm Scheele, Johann Wilhelm Ritter, William Hyde Wollaston, Thomas Johann Seebeck, and Edmond Becquerel. See Josef Maria Eder, *History of Photography* (New York: Dover Publications, 1978).

53 See Shapiro, *Fits, Passions, and Paroxysms*, 345–47. Unlike Brewster, Herschel did not extract chlorophyll but used "juices of the leaves, stalks, roots, &c. of plants." See Herschel, "On the Chemical Action of the Rays of the Solar Spectrum," 200.

54 Herschel. "On the Chemical Action of the Rays of the Solar Spectrum," 16, and William Hyde Wollaston, "A Method of Examining Refractive and Dispersive Powers, by Prismatic Reflection," *Philosophical Transactions of the Royal Society of London*, 92 (1802): 365–80.

55 M. Edmond Becquerel, "Memoir on the Constitution of the Solar Spectrum. Presented to the Academy of Sciences at the Meeting of the 13th of June, 1842," in Richard Taylor (ed.), *Scientific Memoirs Selected from the Transactions of Foreign Academies of Science and Learned Societies and from Foreign Journals* (London, 1843), 537–57.

56 The other reason Herschel stated was the difficulty of completing the proposed apparatus. See Herschel's letter to the Royal Society, October 19, 1843, MC/3/306, Royal Society Library. See also Schaaf, *Out of the Shadows*, 131–32; and Case, *Making Stars Physical*, 184–85.

8
<hr />

Herschel's Planet

Earth in Cosmic Perspective

> The Isle of Wight is enough to make a geologist out of a stone.
> I had no notion of geology before I saw & thumped this island.
> Now I think I have a good glimmering.
> > John Herschel to William Fitton, September 21, 1831,
> > Friedman Collection

An aspect of John Herschel's science not widely appreciated was his complex view of Earth as a planet and as our entry point for understanding physical and chemical processes everywhere in the universe. Herschel was a crucial and consistent voice in bringing both the astronomy and physics of Newton and the most recent chemistry to bear on the Earth. In this chapter, I provide a view of Herschel's evolving interest in the Earth and in geology. I present his interaction with the Earth sciences as a unifying thread running through his life, pulling together aspects of his work that otherwise may seem disparate activities. I examine Herschel's formation as a geologist and a scientific traveler. Lastly, I discuss some of the ways that Herschel's background in physics, chemistry, and astronomy affected his understanding of geology. Herschel sought to understand the cosmos both by looking up at the stars and down at the Earth.

From Natural Philosophy to Minerals to Crystals

Herschel's education started with his father and aunt, William and Caroline, and a series of tutors and schools to prepare him for university studies at Cambridge (see Chapter 1). The planets, the Sun and

Moon, the Milky Way, nebulae, and comets were family matters in his youth. Earth was a planet circling the Sun. John learned from his uncle Alexander Herschel (1745–1821) and his father practical skills needed for building telescopes: the heating, shaping, and polishing of the metal telescope mirrors and the design and machining of gears, axles, and other parts. From Caroline he received his first lessons in chemistry.

Herschel drew from two very different streams of natural philosophy in his formative years. On the empirical side, he learned from the traditions of British natural philosophers and chemists, such as Joseph Black (1728–99) and Thomas Thomson (1773–1852). Thomson published his *Elements of Chemistry* in 1810, from which Herschel became familiar with John Dalton's atomic theory. Herschel's interests in chemistry, minerals, and crystals began here. An extensive record of Herschel's chemical research is preserved in four experimental notebooks at the Science Museum Archives in London, the first beginning in 1813.[1] Herschel's correspondence and other manuscripts indicate he devoted much attention to chemistry, mineralogy, and crystallography in the 1810s and 1820s.

The second stream of Herschel's natural philosophical education started with Newton's *Principia* and was further developed mainly by French mathematicians in the eighteenth and early nineteenth centuries, especially Laplace and the younger scientists around him (see Chapter 2). During the fall of 1811, Herschel attended lectures on Newton; nearly 200 pages of his notes on Newton are preserved at the University of Texas. Herschel wrote to his father that same month that he had burned over 100 pages of speculations about the "nebular vortex," the formation of the solar system, molecular forces, and comet tails.[2]

Although he also lived in Slough and Cambridge for extended periods in the first decade after graduating from university in 1813, London became the center of Herschel's scientific life. His activities can be pieced together from his letters, diaries, and lab notes during this period, which show that he attended meetings of the Royal Society and the Geological Society; conducted chemical experiments alongside Humphry Davy (1778–1829) and William Hyde Wollaston (1766–1828); assisted Edward Daniel Clarke (1769–1822), professor of mineralogy at Cambridge, in his chemical lectures and analyses;

and repeated experiments he found in the literature. Already at age twenty-two, in 1814, Herschel learned to analyze the chemical composition of minerals and, perhaps because of his close reading of Laplace and Jean-Baptiste Biot (1774–1862), measured refractive indices of crystals, their dispersion characters, and the capillarity of liquids. As he attended Clarke's chemical lectures in 1815, he noted to Charles Babbage that he was "become half a mineralogist – and have been analyzing some specimens for [Clarke]."[3]

We know from Herschel's notes and correspondence that he worked his way carefully through the latest chemical and physical literature while analyzing minerals. Herschel relied on Thomson's *System of Chemistry* (1802) and *Elements of Chemistry* (1810). He was constantly precipitating crystals or buying minerals to tease out their component chemicals. In the 1817 edition of Thomson's *System of Chemistry*, Herschel read that a salt in hot water is "water combined with caloric" (3:117) and that fusion of a metal is a "solution by means of caloric" (3:118). At this time of intense experimentation for Herschel, Thomson detailed methods of crystallization and reviewed the conclusion by French mineralogist René Just Haüy (1743–1822) that "integrant particles" of definite shapes make it likely crystals have definite shapes. Herschel read Haüy's four-volume *Traité* (1801) and heavily annotated the 1801–4 English translation of the two-volume *Analytical Essays towards Promoting the Chemical Knowledge of Mineral Substances* by German chemist Martin Heinrich Klaproth (1743–1817). Thomson also reviewed steps taken by the crystallographer Jean-Baptiste Louis Romé de l'Isle (1736–90) and Swedish chemist Torbern Bergman (1734–84). To understand mineralogical processes in nature, it helped to investigate them under laboratory conditions first. This reading and early analytical experimentation provided Herschel both a broad and contemporary understanding of theory of crystal structure as well as experience of how crystals form, which was closely allied with chemical theory and processes.

Herschel produced "A list of Requisite Tables for Perfecting Chemical Science" during his period of peak chemicizing and mineralogizing. The list laid out an ambitious agenda. It projected twenty-nine tables, some of which would be expected in modern chemistry such as lists of simple substances and compounds, chemical density, and freezing and melting points. Other tables used older language and concepts taken from Thomson and others ("caloric absorbed in melting & vaporization"). Herschel included both crystallography and optics in the list: indices of refraction and double refraction, dispersion, and capillary action were included as well as lists of crystalline forms. Herschel clearly aimed to understand the intimate conditions of matter, and he used all the experimental and theoretical tools available.[4]

As preparation for mineral and crystallographic research, Herschel extracted twenty-seven pages of data from Biot's 1816 *Traité de physique*, including information on gas expansion, water density versus temperature, capillarity formulae, refractive powers, and "Double Refracting Chrystals [*sic*]". This data partly met the needs set out by his "Requisite Tables." These notes also indicated the kind of research Herschel envisioned. This was natural history to the extent that Herschel constantly sought new samples and applied new techniques. But it was natural philosophy too, in that Herschel was seeking more general understanding both via enumeration and by taking a quantitative approach. In the sense that natural history and natural philosophy both included broad ranges of investigations, Herschel felt little need to restrict his sense of what is physical and what is chemical.[5] He was not alone in this sense that chemical and physical phenomena are part of one science, but he later articulated his view of natural philosophy more fully than others in his 1830 *Preliminary Discourse on the Study of Natural Philosophy*.

From Mineralogy to Geology

In the context of minerals and crystals, Herschel began interacting with geologists, especially members of the Geological Society. The Society, established in 1807, was growing rapidly at this time – from an original thirteen members to four hundred by 1818. Herschel frequently corresponded on minerals and crystals with Clarke, Davy, and Wollaston, all early members of the Society. Other correspondents on minerals and crystals among his Cambridge friends included especially Charles Babbage and William Whewell. Herschel wrote Whewell in 1815 that the Society was excavating near Slough:

> Extensive subterranean excavations have been making [*sic*] in our neighbourhood undertaken I understand by the Geological Society for

the purpose of attaining an exact and scientific knowledge (not drawn from vague reports of travellers etc.) of the various strata which compose the surface of our globe, & some insight into its internal structure.[6]

After many months grinding and polishing telescope mirrors, Herschel "mineralogized" across the Devonshire countryside for a month in 1817 with Babbage and another college friend. Herschel had to withdraw from a second mineral excursion to Cornwall planned for 1818 but instead experimented on crystals at home at Slough. Herschel worked his way into the network of mineral collectors and providers both in London and abroad, including Wilson Lowry (1762–1824, an early member of the Geological Society and husband of mineralogist Rebekah Delvalle, 1761–1848) and Samuel Young. For the kinds of chemical and optical investigations of crystals that Herschel wished to make, quality crystals were needed, sometimes in quantity.

Herschel's transition from chemical analysis of minerals to the interaction of polarized light with crystals provided a new tool for studying crystalline structure. This concern with the intimate structure of crystals persisted throughout Herschel's career alongside an interest in the formation of crystals and minerals during chemical interactions or in fusion or solution. From 1818 to the early 1820s, as Whewell wrote, Herschel was "untwisting light like whipcord," using polarized light to probe crystals. Herschel verified Huygens' law of the refraction of the extraordinary ray in Iceland Spar. He discovered that the axes of crystals differ for different colors of light. Finally, he showed that a structural asymmetry in many quartz crystals is associated with rotation of the plane of polarization of light transmitted along the crystal's axis. Alongside David Brewster (1781–1868) and Biot, Herschel combined the tools and theories of crystallography (goniometers, integrant molecules, etc.) with those of light rays (changes in refraction and polarization) interacting with crystals. Herschel's library included publications on the relevant optics by Brewster and Biot and, on the crystalline side, the 1822 article "On the Determination of Certain Secondary Faces in Crystals" by the French mathematician and mineralogist Armand Lévy (1795–1841). Herschel interacted closely with Whewell regarding his ideas on representing crystalline structure mathematically. A few years later, in 1827, Herschel finished his long article

"Light" for the *Encyclopedia Metropolitana*, in which he reconciled crystalline optics with the wave theory of light. These discoveries regarding crystalline structure survived the transition in optical theory from viewing light as corpuscles to waves.[7]

Herschel circulated his first mineralogical paper, on the nature of "Swedish feldspar," to a few members of the Geological Society between 1814 and 1816. In 1816, he learned that his research on the components of feldspar trespassed on territory well-trod by Swedish chemist Jöns Jacob Berzelius (1779–1848) and English chemists Wollaston, Smithson Tennant (1761–1815), and James Smithson (1765–1829), who later endowed the Smithsonian Institution.[8] Herschel withdrew this paper. However, in the process he learned about aluminum silicates involving potassium, barium, and other alkali metals. These minerals are richly represented in volcanic rocks, in which Herschel later took an interest. As Herschel recorded in a chemical notebook, the "complicated circumstances under which the formation of stones must have taken place" in nature far outweigh the conditions of the laboratory, "even with every artificial aid, to insure (in the Laboratory) the perfect purity & uniformity of our results."[9] Herschel was beginning to conceptualize the formation of rocks in nature as he investigated rocks and minerals in the lab.

Herschel's publications early in his career did not reflect this broader interest in the Earth and its geological processes. After his first attempt at a geological publication in 1816 up to 1820, Herschel paid little attention to current questions in geology distinct from mineralogy. All of his articles in the *Philosophical Transactions* were mathematical before 1819. He did not publish explicitly about geology until 1825. His letters and notes say nothing about fossils or rock types before 1821, although he likely knew about discussions of strata and debates about the roles of fire and water in producing rocks. To find the first detailed evidence of Herschel's increasing interest in geology, one needs to look at his correspondence, diaries, and observational notebooks.

What geological resources did Herschel have available? Thanks to his father and aunt, Herschel grew up with a rich scientific library.[10] This included the 1785 and 1788 installments of *Theory of the Earth* by Scottish geologist James Hutton (1726–97), which Hutton had presented to William, and *Illustrations of the Huttonian Theory of the Earth* (1802) by John Playfair (1748–1819). The library also held *Mineralogy* (1815) by English chemist Arthur Aiken (1773–1854) and a few related articles. Most importantly, the Herschel family library included the *Transactions* of the Geological Society from volume one (1811), two volumes of the second series of the *Transactions* (1822 and 1824), the *Abstracts of Meetings* (Numbers from 1 to 35), and the *Proceedings* from volume one (1826 ff). Included in John Herschel's collection of separately printed articles (author's copies of articles not bound in the journal) was James Hall's "Account of a Series of Experiments Showing the Effects of Compression on the Action of Heat" (1812) and "On the Consolidation of the Strata of the Earth" (1826). The Herschel library also held books on stratigraphy and geological travel, such as *Geological Travels in Some Parts of France, Switzerland, and Germany* (1813) by Jean-André De Luc (1727–1817), *Outlines of Geology of England and Wales* (1822) by William Daniel Conybeare (1787–1857) and William Phillips (1775–1828), and "On the Physical Structure of those Formations which are Immediately Associated with the Primitive Ridge of Devonshire and Cornwall" (presented in 1820, published in 1822) by Adam Sedgwick (1785–1873). This range of important geological publications would have been available to Herschel from the very beginning of his transition from crystallographic lab work to geological field work.

Herschel's exposure to geology increased in 1820, with he and Babbage planning another trip to mineralogize in Cornwall.[11] Herschel began corresponding more frequently about planning trips to geologize, especially as he deepened his connections to geologists and the Geological Society. The first documentation I have found of Herschel attending a meeting of the Geological Society is upon his election as a fellow in 1824, but Herschel's close association with Clarke, Wollaston, and Lowry likely brought him to Society meetings prior to then. Herschel's other contact point with geologists was through his college connections. In 1820, Herschel corresponded with Sedgwick about the fledgling Cambridge Philosophical Society. On Monday, April 17, 1820, Herschel presented a paper on plagiedral quartz to the Cambridge society. The next evening Sedgwick invited him to dine "in Hall" at Trinity College, and on Wednesday Sedgwick showed Herschel the Woodwardian collection of minerals. Herschel rushed to London for some essential business but was back in Cambridge by Saturday the 22nd for dinner with, among others, Sedgwick and William Buckland (1784–1856), who had been appointed reader in geology at Oxford University in 1819.[12] Herschel and these geologists may have discussed Sedgwick's geological field work conducted in 1819 in Devonshire and Cornwall, which Sedgwick had presented to the Cambridge Philosophical Society in March. That summer, Herschel visited Devonshire again to geologize with Babbage and may have made his first geological visit to the Isle of Wight.[13] Sedgwick later that year wrote Herschel about strata of oolite and mineral specimens. From this point on, Herschel became increasingly active among the geologists and familiar with both field work and lab-based research.

William Whewell, Herschel's close friend and fellow Cambridge graduate, provided Herschel with opportunities for discussion of crystals and minerals and the investigation of their structures and compositions. Their friendship spanned undergraduate discussions of mathematics and John Locke to, in later years, debate over reform at the University of Cambridge (see Chapter 10). In 1818, Herschel reported to Whewell on the structure of nitre and Iceland spar. In the 1820s, they corresponded about Whewell's system of notation for crystal faces.[14] Whewell's approach provided an analytical representation of crystalline structure, which suggested to Herschel the "primitive molecules" comprising the crystal.

Herschel's interest in crystals did not abate as his interest in field science grew. In 1820, Babbage arranged for Lowry to cut and polish calcite crystals destined for Herschel's optical analysis. Herschel received from his St. John's College friend James Gordon "a set of crystals of quartz cut & [polished] and verified by their means my conjecture on the connexion of the phaenomena of rotation with the cause producing the unsymmetrical faces in the plagiedral variety of Hauy [sic]."[15] But Herschel was now complementing his strengths in laboratory studies of minerals and crystals by learning about strata, rock types, and fossils from geologists who gathered their data in the field. He soon began in his own work to bridge theory, laboratory, and field.

Philosopher Abroad: Field Work in the Alps and Beyond

The 1820s represent a distinct stage in Herschel's scientific life. He was beginning to imagine a phase in the development of science beyond the work of individual researchers, a new approach that would require the collaboration of many observers and mathematicians and support across national boundaries. He saw this as a fruitful approach for astronomy but also began to see its necessity for geology and other Earth sciences. The scope of global sciences and the time dimensions of earthly processes demanded quantitative research, standardization of instruments, and international organization. As in the case of double stars and nebulae, Herschel saw the Earth changing over long time periods, thus requiring standards and quantified measurements. An in-depth examination of Herschel's involvement in the organization of such global activities must await another venue, but from 1820 Herschel definitively embraced and embodied the role of the "Philosophical Traveler" and research coordinator as part of this larger vision of organized, ongoing global science. It is ironic that an astronomer known (erroneously) for working alone promoted scientific enterprises that submerged the individual within anonymous, collective research regimes.

Herschel's name and the fame of his father and aunt, of course, opened many doors in Europe as Herschel pursued his role as global research coordinator and European traveler. A few scientists knew him for his own research: astronomers Carl Friedrich Gauss (1777–1855), Laplace, and Friedrich Bessel (1784–1846) and optical researchers Biot and Augustin-Jean Fresnel (1788–1827), for example. More knew him as Foreign Secretary of the Astronomical Society, as a Royal Society Council member, or as a representative of the Board of Longitude. Herschel used all these roles to his advantage to extend his networks first in terrestrial physics and then in geology.

The 1820s and '30s were Herschel's peak years for international travel and field work. In this role, he built upon a long tradition of philosophical travelers, from Edmond Halley (1656–1742) and Joseph Banks (1743–1820) to Alexander von Humboldt (1769–1859), one of the best-known philosophical travelers of the age. Humboldt stood in high public esteem for his exploration of South and Central America around 1800 and for the romance of his grand project to encompass the cosmos via measurement and standardization. British philosophical travelers participated in the expeditions to the North American Arctic for empire and science by British naval and army officers under John Ross (1777–1856) and William Edward Parry (1790–1855). Herschel met socially with these famous commanders but also with the military engineers and navigators affiliated with them, especially Edward Sabine (1788–1883), whose career became intertwined with Herschel's. Herschel was also inspired by the savants of the nascent Alpine climbing community, in particular Horace-Bénédict de Saussure (1740–99), who pioneered the use of barometers, hygrometers, and other instruments on Alpine peaks and was the third person to reach the summit of Mont Blanc, in 1787. Herschel was young and fit enough to follow these exemplars and make mountain climbing a part of his pursuit of science.

Herschel first toured Europe with research in mind in 1821. He and Charles Babbage arrived in Paris in late July and started the tour with meetings with Laplace, Bouvard, Arago, and Humboldt, among others. Herschel had several objectives during this nearly three-month tour of France, Italy, and Switzerland. First, he was charged by the Council of the Royal Society of London to arrange a British-French collaboration in determining the difference in longitude between the Paris and Greenwich observatories. Astronomical work required a precise knowledge of the geographical location of the observatories. This determination also provided a base for geodetic mapping (triangulation), which was underway not only in France and Britain but in many other countries.[16]

Herschel's personal scientific goal during this European tour was to lay a preliminary basis for international collaboration in an aspect of terrestrial physics. Although most observatories and many scientists had barometers, there was at the time no effort to standardize their readings. This prevented useful comparisons and gave the illusion of meaningful geographically distributed measurement. Herschel and Babbage each carried a new mountain barometer, made by the respected London firm of Troughton and Simms, on their journey. They planned to conduct comparative measurements of their barometers against those at the Paris Observatory and all others they visited in France, Switzerland, and Italy. They also intended to test the utility of barometers in measuring the altitudes of mountain peaks. To this end, they organized several coordinated measurements to be made in towns while they carried their barometers up mountainsides to make simultaneous measurements at summits and other points of ascent. Herschel kept detailed notebooks of these measurements and clearly had in mind the disciplining of individual observation and data handling required for long-term, collaborative Earth science.[17] Herschel ascended the Breithorn on September 7, 1821, which his guide mistook for Monte Rosa, the second highest mountain in the Alps.[18] Geodesy and barometry – both topics of terrestrial physics – dominated Herschel's attention on his first extensive exploration.

Geology was a secondary interest during the 1821 trip; nevertheless, Herschel paused dozens of times in his travels to sketch mountain scenes with his camera lucida. He also kept an eye out for crystals and minerals. Herschel's description of Mont Blanc to his mother from Chamonix of August 15, 1821 emphasized the romantic aspect of his first scientific tour of the Alps:

> Nothing can be conceived more beautiful than the mild light as it falls on the rounded dome which forms the top of this enormous mass, and the numberless pinnacles, sharp as needles and shooting up in pyramidal groups round its base, all splintered in a thousand rifts, and covered with snow wherever the steepness of the rock will suffer it to rest. Immediately in front of my window is a huge hill probably 5 or 6 thousand feet high, covered from top to bottom with a dark forest of pines, and which forms the immediate base from which the loftier peaks spring. The *Aiguille de Midi*, which seems so close one might almost touch it is at least 12000 feet high. The little scratch [sketch] at the corner may give you some idea of the sharpness and fractured form of the granite rock which it consists of.[19]

Here in the cradle of European mountaineering, where Saussure, Marc-August Pictet (1752–1825), and others did so much in the 1770s to 1790s to develop climbing and the pursuit of science at altitude, Herschel came to terms with the imposing scale of mountains. He explored the human capital of mountaineering: climbing guides and local sellers of crystals. He and Babbage visited glaciers and rode up on mule-back several thousand feet in elevation. Among other sites, they reached "The Garden," reputedly the highest growth of grasses in Europe, by the Mer de Glace glacier. Their guide, Joseph-Marie Couttet (or Coutet), claimed to be the son of Marie Couttet, who accompanied Saussure in his 1787 summit climb of Mont Blanc. As Herschel wrote his mother, Couttet was cautious or bold as the situation demanded and understood both geology and mineralogy. Though they ascended its slopes, Herschel expressed no interest in climbing to the summit of Mont Blanc. This reflected his disinterest in first ascents and the higher importance he gave to the research to be conducted. Perhaps, though, he also realized he was not prepared for the magnitude of the highest peak in the Alps.

During the 1821 excursion, Herschel sought out crystals he could use in experiments with polarized light. He was especially interested in plagiedral quartz crystals, which rotate the plane of polarization of light and on which he had published an article in the *Philosophical Transactions* in 1820. As Herschel wrote to Brewster:

> The law of rotation in Plagiedral Crystals, which rested on too slight an induction, I have lately verified in 30 additional specimens of very fine Crystals of this variety which I brought with me from Mt. Blanc (where the crystals have almost all this peculiarity) & have encountered no instance of an exception. It is rather remarkable that left-handed crystals seem to be there, much more frequent than right – at least among a great number I examined their proportion was nearly as 2:1.[20]

Herschel made only a few geological observations on this first Alpine journey. He sketched the contorted strata of Mt. St. Julien, near Modane, France, and of the Mont Cenis pass into Italy.[21] He also drew the Jura Pass and the Simplon Pass, a glacier at Zermatt, and the clay columns near Stalden, Switzerland. His interest in geology, however, was soon to increase.

Before Herschel returned to the mountains in Europe in 1824, he immersed himself more fully in geology as well as following up on his research with crystals and polarized light. He noted in his diary what appears to be his first meeting with Charles Lyell, on February 9, 1822, and on February 28, he recorded: "Dined with Club. Had a big confab with Buckland."[22] Buckland, who Herschel had known for at least two years, had become a celebrity after his discoveries of hyena bones in caves in England. In 1823, Herschel was asking Lowry for help identifying a piece of granular quartz and promising to show him some garnets obtained from John Stevens Henslow (1796–1861), who found them in Anglesea and was just beginning to undertake fieldwork himself.[23] Two months before Herschel began his second and more extensive European tour in 1824, he was admitted as a Fellow of the Geological Society at a meeting presided over by Buckland.[24] A few weeks later, he dined at the Club of the Geological Society and attended a presentation of William Conybeare's Plesiosaurus. Herschel reported a large crowd was in attendance and that "Buckland [was] very facetious in the chair."[25] Herschel's geological network was coming together.

Herschel, drawing in part on the influence of his new network of associates, had a new geological focus on his second, six-and-a-half-month European excursion: volcanoes. This time he traveled with only his servant, James Child. He began this tour again in Paris but stayed only three days, making sure nonetheless to see Humboldt, Laplace, Arago, and Fourier. He wrote to his mother that Humboldt gave "useful information on the Alps and Vesuvius."[26] In particular, Humboldt passed Herschel an article by Christian Leopold von Buch (1774–1853), "Géologie du Tyrol," published in 1822. According to Sydney Ross, Herschel's copy of this article is full of marginalia: "Examined on the spot. Found correct." Herschel's scientific notetaking during this extended tour indicated increased attention to geology. He filled four travel journals, which are in the Herschel Family Papers in Texas.[27]

Herschel did not visit Geneva on this trip but went via Lyons directly to Mt. Cenis and the road into the Susa Valley and Turin. Leaving Paris on April 7, he made it to Mont Cenis by the 14th, where he entered measurements made at the Napoleonic military barracks, the main shelter in the mountain pass, with barometer, thermometer, and hygrometer. He was keenly aware of the intense heat from the Sun at these altitudes, perhaps recalling Babbage's severe sunburns from their earlier trip. This inspired him to design and build his first "actinometer," an instrument for measuring the intensity of the Sun's rays, a week later when he stayed in Turin to meet with the astronomer Giovanni Plana (1781–1864).

Herschel finally arrived in Naples for the geological/volcanological portion of his tour in early June. A few days later, Herschel, James, and a small party set off at evening to climb Vesuvius and reach the summit at sunrise. Herschel wrote to his mother that they passed along the lava flow of 1810 and tied up their burros at the beginning of a steep climb to the top. He "made haste to climb to the Edge" and was astounded at the profound depth of the crater. While his associate the Italian mineralogist Nicola Covelli (1790–1829) collected gases at a fumarole and performed chemical experiments, Herschel straddled the lip of the crater and went partway around to a spot where he could set up his barometer and other instruments. He also made one of his usual spectacular camera lucida drawings.[28]

For Herschel, volcanoes provided access to materials and structures deep within the Earth. He described this volcanic structure as much bigger than Vesuvius, stretching from Etna to Bologna along the western slope of the Apennines. This "vast volcanic region" had existed, Herschel ascertained, since antiquity. The power of this system was evidenced in the strata of lava ninety feet thick and in the hot baths and steam still found there. Herschel was hatching a plan to study active volcanos using the best physical and chemical analytical techniques. The visit to Vesuvius was relatively easy and allowed him to warm up for Mt. Etna. Having heard a report of Sicily from an Englishman he met, Herschel found it was simple to get a fast steamer to Palermo in Sicily, and he and James were off. In Sicily, Herschel could prepare to climb and study one of the highest active volcanos in Europe.

To climb Etna required assistance. Herschel brought with him letters from both Italian professors and the British Proconsul in Naples to smooth the way. He especially needed the help of Mario Gemmellaro (1773–1839), "who is to Etna what [Teodoro] Monticelli [1759–1845] is to Vesuvius, its active & scientific investigator, and the recorder of all its Phaenomena."[29] As always for a visiting metropolitan scientist, these local volcano experts were essential. Herschel found Gemmellaro two weeks into the trip when he finally reached Catania, at the base of Etna.

Herschel also wanted to traverse the island for geological observations on his way to Mt. Etna, a distance of over three hundred miles before even reaching the mountain. He rode a mule in a mule train since there were few cart roads on the rocky, desolate, and isolated island. He wrote to his mother that at the sulphur mines at Catolica, "I loaded myself with minerals till the muleteer swore it was impossible for the mules to carry them."[30] He lamented that he found mineralogical riches he felt compelled to collect despite the heat of the Sicilian sun:

> in the midst of the ancient lavas of the extraordinary volcanic district surrounding Etna, and I had no choice but to give up the examination of one of the most remarkable features of the country, or encounter noonday Sun. Therefore off I trudged with my hammer and bag and was amply rewarded for a good broiling by some of the best minerals I have found.[31]

Herschel had three goals in climbing Mt. Etna. The first was to observe the historical lava flows and study the rocks and gases of the volcano. The second and third goals treated Mt. Etna as an access ramp to high altitude: more careful barometrical measurement for calculating the elevation of the mountain and further testing of Herschel's new actinometer for measuring solar radiation incident on Earth. On the night of July 3–4, Herschel's party (two guides, James, and himself) spent the night at the Casa Inglese, a rough stone shelter Gemmellaro had built for guiding climbers. While the others snored, Herschel had so many measurements under way he could not rest. With the barometer measurement of 21.3 inches, he placed the altitude of the Casa at 9,590.7 feet above sea level, leaving about 1,500 feet to ascend next morning.

The final ascent to Etna's summit the next morning, Herschel later recounted, was "most fatiguing" – and no wonder since, as he wrote, he "got some excellent observations" during the night. His fatigue was compounded by the thin air and by the difficulty climbing in ash and cinders.[32] Despite his weariness, Herschel was impressed at the difference in scale between Vesuvius and Etna:

> The vulcano [sic] is upon a scale far exceeding Vesuvius, being in fact nearly three times the height, and is extremely active. The vestiges of its fury are visible in all directions. All the upper part of the mountain is composed of ashes, scoria and ragged lava, the middle region is thickly wooded with pine and old oak and green with fern, putting me much in mind of many parts of Windsor Forest, but the woods are intersected in every direction by vast black rugged mounds, which at a distance look

out above the trees like piles of fresh earth recently thrown up from some great excavation, but on a nearer approach are found to be composed of huge masses of excessively hard shapeless & vitrified stone. These are the lavas of successive eruptions which have burst from craters at different heights on the flanks of the mountain and rushed down burning their way through the forests and precipitating themselves on the vineyards & towns below.[33]

The company he found in Catania provides an indicator of Herschel's self-image on this trip. From Catania, he arranged to return via an overland route to Palermo with Count Beffa Negrini from Mantua, who "had a large hammer & bag." In Catania he also met Charles Daubeny (1795–1867) of Oxford University and planned to see him again in Naples and geologize together. Herschel noted that Daubeny also had the requisite hammer and bag. Herschel's identity as a geologist was strengthening.

Herschel described a variety of geological features as he traveled overland from Catania to Palermo. High up Etna were the forty-seven-year-old lava rivulets, dating from an eruption, frozen in time. He noted a high pyramidal hill of sandstone atop volcanic rock. In generalizing about how salts are deposited, he noted that in Sicily as elsewhere, they often lie under a bed of clay and above a bed of gypsum. Returning to Naples, he shipped a large amount of minerals collected in Sicily home to Slough. His journey then continued on into the Tyrol and Germany, with Herschel collecting minerals and describing geological features along the way. He made good use of von Buch's geological report on the Tyrol, and he worked with mineral suppliers and guides to seek out particular mineral and bone deposits. One sketch in Travel Journal VII is of peaks overlooking the valley of Fassa in Tyrol. It shows hill after hill with annotations: Calc, Pyrox, Dol, and "hard grey [illegible] in larger broken nodules. Grit Pyrox! nodal concretion not hard rock."[34] Herschel's familiarity with minerals and rock types rivaled his knowledge of the night sky.

Herschel's geological field work continued after this extensive trip, with excursions in 1826 to the volcanic districts of Auvergne and Vivarais, in 1827 to North Wales and to the Giant's Causeway in Ireland, and to the Isle of Wight in 1828 and 1831. These trips included "grubbing for fossils," as Herschel called it. His geological work also continued during his sojourn near Cape Town in the 1830s.[35] The details of these trips as well as Herschel's participation in the Geological Society and interaction with well-known geologists remain insufficiently examined. Many historical problems lie waiting to be addressed in this material, and few scholars seem to be aware of them. Part of the reason may be that it is hard to imagine someone so well-known as an astronomer taking such an interest in the Earth. In the next section, I turn from geology to examine how Herschel's intimate involvement in astronomy affected his perspective on the Earth.

A Planet in the Cosmos

It may seem obvious to state that John Herschel accepted the Copernican heliocentric explanation the planetary system, but it bears emphasis because Herschel took extraordinary care to present Earth in this unifying, natural philosophical perspective. In his classic *Treatise on Astronomy* (1833), Herschel affirmed:

> We shall take for granted, from the outset, the Copernican system of the world; relying on the easy, obvious, and natural explanation it affords of all the phenomena as they come to be described.[36]

Herschel admitted that readers might find it strange that a book on astronomy included Earth among the planets, a heavenly body.[37]

> Thus, the earth on which he stands, and which has served for ages as the unshaken foundation of the firmest structures, either of art or nature, is divested by the astronomer of its attribute of fixity, and conceived by him as turning swiftly on its centre, and at the same time moving onwards through space with great rapidity.[38]

Indeed, this astronomical text included a first chapter of fifty-five pages arguing in favor of a spherical Earth and movement without sensation, providing images of Earth's dimensions and surface. Herschel was careful to emphasize the scales involved. Though the oceans were as deep as the highest mountains, they were as a mere film on the globe of the Earth. The atmosphere, Herschel wrote, effectively ended at about one-hundredth part of Earth's diameter, or about eighty miles above Earth's surface. "We are thus led to regard the atmosphere of air, with the clouds it supports, as constituting a coating of equable . . . thickness, enveloping our globe on all sides, or rather as an aërial ocean." Astronomers needed to account for atmospheric refraction due to the successive "strata" of air to correct stellar positions. And the minuscule measure of stellar parallax indicates that Earth – large as it seems – is extremely small on the cosmic scale.[39] Herschel was slowly building the basis for appreciating Earth as an astronomical object, a planet like other planets.

Earth also figured prominently in chapter 3, "Geography."[40] Because all astronomical observations were made from Earth's surface, Herschel wrote, a knowledge of that surface was important to astronomers. The part of geography relevant to astronomy related to exact determination of the surface features of latitude, longitude, and elevation. Herschel discussed the measurement of a degree of latitude along a meridian. On comparing lengths of a degree reported around the globe, Herschel noted that the measured length of a degree is greater nearer the poles. In this, Herschel was introducing for his general audience geodetic surveying and Earth's ellipsoidal, not spherical, shape.[41]

Herschel also gave attention to topics he saw as beyond astronomical geography but part of physical geography. These sometimes approached geology. When Herschel turned from observation of Earth's ellipsoidal shape to theory, for example, he quipped that it was a good thing Earth turned out to be this form, since this "is precisely what *ought theoretically* to result from the rotation of the earth on its axis."[42] To illustrate, he proposed a thought experiment of an Earth-like globe covered in water and set gradually to rotating. Centrifugal force would cause water to flow away from the poles (and the axis of rotation) and toward the equator. Herschel continued his illustration, pointing out that the astronomical rotation of Earth would have geological consequences. Over the long history of Earth, the sea beat on the land, "grinding it down" and conveyed it to ocean depths:

> Geological facts afford abundant proof that the existing continents have all of them undergone this process, even more than once, and been entirely torn in fragments, or reduced to powder, and submerged and reconstructed. Land, in this view of the subject, loses its attribute of fixity.[43]

Herschel extended Copernicus' idea of a mobile Earth to the geological features of its surface. Astronomical time and geological time were both immensely long, whereas Earth's surface was transitory. Herschel deduced this was from the balance of gravity and centrifugal force, though he denied "meaning here to trace the process by which the earth really assumed its actual form."[44] It was enough that observed geology was consistent with expectations from natural philosophy. Nevertheless, it was noteworthy that although the surface, the geological features of Earth, changed, Earth's form achieved balance as an oblate spheroid.

Herschel next calculated the ratio of the centrifugal force at Earth's equator to the weight of a body to be 1/289. He advised readers that, in fact, precision measurements of gravity had been made in different latitudes and found to conform roughly with theory but that the value measured was 1/194. (He here referred to Edward Sabine's pendulum studies in the arctic and tropics in the 1820s.)[45] This difference was due to the elliptical form of Earth. That is,

> owing to the elliptical form of the earth alone, and independent of the centrifugal force, its attraction ought to increase the weight of a body in going from the equator to the pole by almost exactly 1/590th part; which, together with the 1/289th due to centrifugal force, make up the whole quantity, 1/194th, observed.[46]

Herschel next discussed consequences of Earth's physical geography: Trade Winds, the exposure of Earth's surface to solar radiation, the different weights of fluids of different temperatures, and the effect of Earth's rotation on atmospheric movements. The physics that drove the Trade Winds, Herschel suggested, might also explain hurricanes.[47] Astronomy and geography came closer together in the burgeoning precision mapping that was underway in the nineteenth century. Herschel devoted a dozen pages to various means of measuring latitude and longitude at any station on Earth.[48] From this, he turned to the use of theodolites and triangulation on a base line. He noted that in fact the triangles measured are ellipsoidal triangles, not flat or spherical ones, and that the difference is observable when triangles are six or seven miles long.[49]

Like others before him, Herschel had observed that Earth is made of two very different hemispheres: one containing nearly all the land and the other nearly all the ocean. This differentiation was massive enough, he wrote, to qualify as "an astronomical feature."[50] Not only would it be visible from space, but it would affect Earth's interaction with other astronomical bodies, especially the Moon. Herschel tried to understand this feature in terms of the physics of the Earth. Since the whole globe – water and land – is in equilibrium, the land protruding must be buoyant with respect to the rock beneath it. Herschel expressed here an idea he returned to many times: that the buoyancy of continents must be accounted for in geological theory.

> We leave to geologists to draw from these premises their own conclusions ... as to the internal constitution of the globe, and the immediate nature of the forces which sustain its continents at their actual elevation.[51]

Herschel's ideas about buoyancy, heat flow, and Earth's deep structure occupied the boundary between geology and physics of the Earth.

The Sun was, in Herschel's view, the primary cosmic influence on terrestrial phenomena. It alone of celestial bodies acted on Earth beyond gravitational attraction. Solar radiation, he wrote, produced most motions on Earth, including the winds and ocean currents. The Sun produced disturbances in "the electrical equilibrium of the atmosphere," thus causing phenomena of terrestrial magnetism. The Sun was responsible for plants, the cycle of evaporation and precipitation, and coal. The Sun was even responsible for geological change since it drove the wind and rain that slowly degraded solid rock and carried it out to sea.[52]

Herschel saw Earth in so strong a Copernican light that he visualized it as a planet seen from space. The law of gravity and laws of motion governed its orbit and its shape, but there was more to the physics of Herschel's cosmos than mechanical forces. Likewise, there was more to the physics and chemistry of Earth. What caused volcanic regions or the rise of mountains or continents? What caused oceanic currents and tides, weather systems and storms, auroras and Earth's magnetic phenomena? Herschel saw not only geology but also meteorology, geodesy, and terrestrial magnetism as all ultimately part of one dynamic science of the Earth.

In conclusion, John Herschel was an enthusiastic, well-informed geologist who immersed himself in mineralogical research, then in geological field work, and ultimately attempted to reconcile the evidence of geology with the requirements of natural philosophy. I position Herschel's early interest in geology in the context of his chemical and mineralogical research. Hence, when Herschel did turn to geological fieldwork, he saw strata and mountains in terms of mineral composition and gradual change. His growing connection to the geological community introduced him to issues of strata, genesis of rocks and minerals, and ideas about catastrophism and uniformitarianism, Neptunism and Vulcanism, and the deep well of time indicated by new evidence. Likewise, his connection to other communities – astronomers, alpinist-savants, arctic research officers – broadened Herschel's developing notion of terrestrial physics: how physical and chemical principles applicable in the cosmos also underlay all earthly processes. Herschel conceptually unified the Earth with the cosmos in both space and time by insisting on uniform, quantitative observations, rigorous analysis, and the universal law of the natural philosopher's cosmos. He expected that all observations of Earth – geological, meteorological, geophysical – had ultimately to be consistent with each other. There was but one science.

Notes

1 John Herschel, "Experiments in various subjects viz optical, chemical and nonsensical and queer things miscellaneously arranged for the benefit of posterity," four manuscript notebooks. Science Museum Archive, London, MS/0478.
2 John Herschel, "Notes taken by John Herschel at Cambridge while attending lectures on Newton's Principia, 27 September 1811 and 3 November 1812," University of Texas, Humanities Research Center, Herschel Family Papers (hereafter HRC), containers 8.9 and 7.19, respectively. Also, John Herschel to William Herschel, November 16, 1812, container 26.4.
3 John Herschel to Charles Babbage, March 23, 1815, RS:HS 2.38.
4 John Herschel, "A list of Requisite Tables for perfecting Chemical Science," no date, HRC, container 3.11.
5 John Herschel, "Memoranda, data, etc. from Biot's *Traité de physique*," 1816 or later, HRC, container 11.46.
6 John Herschel to William Whewell, February 4, 1815, Trinity College, Cambridge, Camb. Add. Ms.a.207[1].

7 See Gregory A. Good, "John Herschel's Optical Researches and the Development of His Ideas on Method and Causality," *Studies in the History and Philosophy of Science*, 18 (1987): 1–41.

8 Henry Warburton to John Herschel, April 24, 1816, RS:HS 18.36.

9 Herschel, "Experiments in various subjects." In the first notebook, following Experiment 217, Herschel composed two drafts of his experiments on feldspar.

10 Sydney Ross (ed.), *The Catalog of the Herschel Library* (Troy: Privately printed, 2001).

11 Herschel to Babbage, April 12, 1820, RS:HS 20.88.

12 Herschel, diary 1820, entries for April 18, 19, 22, and 30, 1820; diary 1822, entry for February 28, 1822, HRC, container 16.7.

13 Herschel to Babbage, July 12, 1820, RS:HS 2.138 and Herschel, "Painting No. 6, Copied from Camera Lucida drawing No. 538, Cliffs on the Main Beach, Isle of Wight."

14 Herschel to Whewell, no date [October 15, 1823 or later], TC, Camb. Add. Ms.a.2078.

15 Herschel to Babbage, January 5, 1820, RS:HS 2.124 and Herschel, diary 1820, entry for March 3, HRC, container 16.7.

16 See, for example: James Lequeux, "Geodetic Arcs, Pendulums, and the Shape of the Earth," *Journal of Astronomical History and Heritage*, 23.2 (2020): 297–326.

17 Gregory A. Good, "A Shift of View: Meteorology in John Herschel's Terrestrial Physics," in J. R. Fleming, V. Jankovic, and D. R. Coen (eds.), *Intimate Universality: Local and Global Themes in the History of Weather and Climate* (New York: Science History, 2006), 35–67; and "The Astronomers Who Fell to Earth: Or, How the Copernican Revolution was Completed in the Alps," *Physis: Rivista Internazionale di Storia della Scienza*, New Series, 56 (2021): 7–19.

18 Gregory A. Good, "John Herschel's Travels through the Alps to the Cosmos in the 1820s," in Fabio D'Angelo (ed.), *The Scientific Dialogue Linking America, Asia and Europe between the 12th and the 20th Century: Theories and Techniques Travelling in Space and Time* (Naples: Associazione culturale Viaggiatori, 2018): 288–91.

19 John Herschel to Mary Pitt Herschel, August 15, 1821. HRC, TxU L-0517.3, container 25.14.

20 John Herschel to David Brewster, no date (after October 1821), RS: HS 4.272.

21 John Herschel, J. Paul Getty Museum, Herschel Collection, "Contorted strata of Mt. St. Julien near Modane. 22 August 1821." CL347. Getty Institute LA 91.GG.98.68.

22 Herschel, diary 1822, entries for February 9 and 28, 1822, HRC, container 16.9.

23 Wilson Lowry to John Herschel, April 25, 1823, RS:HS 11.492.

24 John Herschel to Mary Pitt Herschel, February 4, 1824, HRC, TxU:H/L-0517.2 and -0517.3, container 25.14.

25 Herschel Diary 1824, entry for February 20, HRC, container 16.11.

26 John Herschel to Mary Pitt Herschel, April 7, 1824. HRC, TxU L-0517.5, container 26.1.

27 Herschel, Travel Journals IV, V, VI, VII, and VIII, HRC, container 22.5 through 22.8. The location of Herschel's copy of von Buch's article is not known. It would provide detailed understanding of how Herschel's geological ideas were changing.

28 John Herschel to Mary Pitt Herschel, June 8, 1824. HRC, TxU L-0517,   ' container 26.1.

29 John Herschel to Mary Pitt Herschel, June 20, 1824. HRC, TxU L-0517.13, container 26.1.

30 John Herschel to Mary Pitt Herschel, June 27, 1824. HRC, TxU L-0517.14, container 26.1.

31 John Herschel to Mary Pitt Herschel, July 12, 1824. HRC, TxU L-0516, container 25.13.

32 John Herschel to Mary Pitt Herschel, July 1, 1824. HRC, TxU L-0517.15, container 26.1.

33 John Herschel to Mary Pitt Herschel, July 12, 1824. HRC, TxU L-0516, container 25.13.

34 Herschel, Travel Journal VII, HRC, container 22.8.

35 Gregory A. Good, "John Herschel's Geology: The Cape of Good Hope in the 1830s," in Jed Buchwald and Larry Stewart (eds.), *The Romance of Science: Essays in Honour of Trevor H. Levere*, (Cham: Springer, 2017): 135–50.

36 Herschel, *Treatise on Astronomy* (London: Longmann, 1833), 4.

37 Herschel, *Treatise on Astronomy*, 10.

38 Herschel, *Treatise on Astronomy*, 2.

39 Herschel, *Treatise on Astronomy*, 9–63.

40 Herschel, *Treatise on Astronomy*, 107–56.

41 Herschel, *Treatise on Astronomy*, 117.

42 Herschel, *Treatise on Astronomy*, 118.

43 Herschel, *Treatise on Astronomy*, 120–21.

44 Herschel, *Treatise on Astronomy*, 121.

45 Herschel, *Treatise on Astronomy*, 123–26.

46 Herschel, *Treatise on Astronomy*, 127–28.

47 Herschel, *Treatise on Astronomy*, 131–32.

48 Herschel, *Treatise on Astronomy*, 132–46.

49 Herschel, *Treatise on Astronomy*, 146–51.

50 Herschel, *Treatise on Astronomy*, 153–54.

51 Herschel, *Treatise on Astronomy*, 149.

52 Herschel, *Treatise on Astronomy*, 211–12.

9

John Herschel and Scientific Standardization

Precise measurement and the production of accurate standards was at the very heart of John Herschel's conception of science. This was not only true of his own endeavours, including those astronomical, but was fundamental to his philosophical vision of scientific practice, as well as his theological understanding of nature and God's role in Creation. Throughout his career, Herschel contributed to several standardizing projects, including the production of units of weight, distance, time, musical pitch, volume, and even the harmonic division of musical notes. In the one full-time position of paid employment that he occupied, as Master of the Mint, he was responsible for regulating the nation's standards of bullion and coinage and, in his role as a trusted scientific advisor to the British government, he promoted standardized practices for global investigations of natural phenomena, including those astronomical, metrological, and magnetical. To establish accepted units of measurement and standardizing regimes brought order to philosophical inquiry, regulation to the nation and empire's economic activity, and organization to diverse societies around the world. In his promotion of worldwide investigations into natural phenomena through standardized observation techniques and disciplined observatories, Herschel ensured that the cultivation of nineteenth-century natural philosophy was inseparable from the wider proliferation of empire: science was a tool of British imperialism, providing knowledge on a global scale and bringing together colonial territories in the pursuit of new understandings of nature. By the time of his death in 1871, few had done more than Herschel to facilitate accurate measurements of nature; and given that precise measurement was increasingly to define scientific practice throughout the nineteenth century, his standardizing efforts were remarkable in their influence. Crucially, however, Herschel was implicit that knowing nature through accurate measures and with standardized units was, in itself, of intrinsic moral value: to count molecules, weigh matter, and record the order of celestial motions was to know God and understand the universe that He had created.

Standardizing Science

Within the philosophical vision of science that Herschel set out in his influential *Preliminary Discourse on the Study of Natural Philosophy*, published in 1830, he argued that the formation of laws depended on number and the act of counting. Herschel declared Number, 'that is to say, integer number, is an object of sense, because we can count; but we can neither weigh, measure, nor form any precise estimate of fractional parts by the unassisted senses'.[1] A natural philosopher could not, he contended, detect minor changes in weight or degrees of illumination, or even the gradual passage of time, through sense alone: these 'matters of quantity' were 'absolutely vague'. The solution was to reduce all such measurement to counting by fixing 'convenient *standards* of weight, dimension, time, &c.' and by inventing

> contrivances for readily and correctly repeating them as often as we please, and counting how often such a standard unit is contained in the thing, be it weight, space, time, or angle, we wish to measure; and if there be a fractional part over, we measure this as a new quantity by aliquot parts of the former standard.[2]

In this way, Herschel identified measurement standards as the crucial means for allowing a scientific practitioner to count natural phenomena: this act being essential to determining the permanent 'great laws of nature'.[3] This was as true for gravity, which relied on calculations of distance and the measured influence of attracting bodies, as it was for chemistry, which was a science of weighing. Indeed, Herschel asserted that it was 'a character of all the higher laws of nature to assume the form of precise *quantitative* statement'.[4] Ideally, any standard should draw its authority from nature, such as the Earth's rotation that provided natural philosophers with a constant standard of time, and should be materially embodied in an object, such as the physical yard or metre. Herschel, however, went much further and argued that the ultimate object of philosophical inquiry was to produce measurements of ever greater precision and from which increasingly accurate numerical statements could be sustained. As he put it,

> it is not merely in preserving us from exaggerated impressions that numerical precision is desirable. It is the very soul of science; and its attainment affords the only criterion, or at least the best, of the truth of theories, and the correctness of experiments.[5]

To know number was to know nature, but this process was contingent on precision, measurement, and standardization.

As much as contemporaries celebrated Herschel's mathematical and philosophical contributions, it was his astronomical work that secured him celebrity in early Victorian society. Taking over his father's vast project to catalogue the nebulous patches and star clusters of the universe, Herschel worked tirelessly through the 1820s and 1830s to map out the stars and bring standardization to astronomical observations, including double star positions and star magnitudes. His 1833 *Treatise on Astronomy* dominated English astronomy for over two decades while his five years at the Cape of Good Hope, from 1833 until 1838, resulted in the recording of 1,708 nebulae, with only 439 having previously been charted. For Herschel, the cataloguing of stars and accurate positioning of celestial bodies was fundamental to examining the universe and its systematic behaviour. Astronomy, in terms of precise measurement, would allow observers to record change over time. While positional astronomy was traditionally a practical concern, producing data that was of use for navigation and time keeping, Herschel conceived of this work as a philosophical project.[6] The determination of the governing laws of nature relied on precise measurements and standardized astronomical practices to build the numerical data on which knowledge of the heavens would be attained.

Herschel's commitment to the standardized collection of data and precise measurement extended beyond astronomical phenomena. On returning from the Cape in 1838, Herschel found himself conscripted into the great scientific crusade of the age: the surveying of the Earth's magnetic properties. That the intensity of the Earth's magnetism fluctuated over time and space, with the north and south magnetic poles constantly changing position, was well known by the early nineteenth century, Portuguese navigators having observed this phenomenon during the late fifteenth century. Yet in the 1820s and 1830s, understanding terrestrial magnetism took on increasing urgency. Influential natural philosophers, notably Alexander von Humboldt (1769–1859), emphasized that the Earth's natural phenomena, such as heat, electricity, and magnetism, were all interconnected and that to examine nature required the study of these forces on a global scale. But the worldwide collection of magnetic data was also important for ensuring safe navigation and securing Britain's increasingly imperialistic commerce: as the globe's magnetism had a varying effect on a ship's compass needle, it was crucial to understand how this phenomenon changed over time and space. Combining philosophical and utilitarian priorities, a group of scientific authorities campaigned throughout the 1830s for the British government to finance a magnetic survey of the world, making use of the nation's overseas territories and extensive naval resources. Uniting members of the Royal Society and the British Association for the Advancement of Science (BAAS), this scientific lobby was keen to get Parliament and the Admiralty to commence this investigation by launching an expedition to the Antarctic, tasked with mapping the magnetic character of the region and locating the south magnetic pole. Under the dominating leadership of Edward Sabine (1788–1883), this campaign continually failed until, in 1838, it succeeded in recruiting Herschel to the cause.[7] Herschel had never made a magnetic measurement in his life but brought unparalleled eminence as a statesman of science, fresh from his Cape triumph, as well as powerful political connections, which included the Secretary of State for the Colonies, Lord Glenelg (1778–1866), the Secretary of the East India Trading Company, James Melvill (1792–1861), the Admiral of the Fleet, the Earl of Minto (1782–1859), the Royal Navy's Hydrographer, Francis Beaufort (1774–1857), as well the Royal Society's president and sixth son of King George III, the Duke of the Sussex (1773–1843). At the same time, Herschel's daughter was a Woman of the Bedchamber to Queen Victoria. After requesting a meeting with the Prime Minister, Lord Melbourne (1779–1848), and the Chancellor of the Exchequer, Thomas Spring Rice (1790–1866), in August 1838, Herschel discussed the prospects of a magnetic survey with Queen Victoria over dinner at Windsor in October. A series of meetings between Melbourne, members of his Cabinet, Herschel, and several leading magnetic lobbyists followed. By early 1839, the government was committed to an Antarctic expedition, which subsequently departed under Captain James Clark Ross' (1800–62) command and would fulfil many of the campaign's ambitions, including the calculation of the south magnetic pole's location in 1841.[8]

Yet Herschel did far more than secure state patronage for this vast magnetic measurement: he fundamentally changed the nature of the project, reconceptualizing it as part of a broader global investigation into natural phenomena in which standardized practices and units would be crucial. To determine physical laws that were universally true required a systematic collection of data and this, Herschel believed, demanded a network of 'physical observatories' that would perform magnetic measurements, as well as record tidal and meteorological phenomena.[9] While Sabine and the magnetic lobbyists emphasized the importance of expeditionary surveys, Herschel wanted the inclusion of a series of fixed observatories, around the world, with standardized observational practices for conducting measurements of extreme precision, including of oceanic tides, terrestrial magnetism, and the weather.[10] When he was at the Cape, Herschel had become convinced of the value of such a network of institutions, having performed hourly meteorological observations in 1835 that he coordinated with similar recordings made around the world, including in Guyana, Mauritius, Australia, and India. The formation of such a system of standardized observatories was the prime inducement for his support of the magnetic campaign, as he made clear to Beaufort in June 1838, calling for the formation of observatories 'over the whole surface of the globe, and especially of establishing permanent magnetic stations at the Cape, in India, Australia and other points within the range of British superintendence'.[11] In this way, he united Britain's scientific and colonial interests, with the magnetic survey representing a rare chance to establish a global system for the measurement of natural phenomena. The observatories that Herschel had done so much to establish ensured that the development of British science was inseparable from the wider expansion of empire, uniting disparate overseas territories more closely with London.

By the time Ross sailed for the Antarctic in the autumn of 1839, Herschel had ensured the state financing of physical observatories at Toronto, Cape Town, Hobart, and St Helena, which would cooperate with foreign observatories throughout Europe, America, Africa, and Russia, as well as East India Company establishments in India and Singapore. This was a standardizing programme of unprecedented scale: to orchestrate coordinated measurements of natural phenomena over vast geographical distances demanded a shared system of practices and values. Indeed, such organization was something Herschel had been acutely aware of in his astronomical work. Eager to remove the necessity of specifying the location of astronomical observations, given that time is contingent on longitudinal meridian, Herschel proposed a standardized astronomical 'equinoctial time'. As his good friend Mary Somerville (1780–1872) put it, Herschel's proposed standard would constitute a universal system for observation that would make time

> the same for all the world, and independent alike of local circumstances and inequalities in the sun's motion. It is the time elapsed from the instant the mean sun enters the mean vernal equinox, and is reckoned in mean solar days and parts of a day.[12]

In other words, Herschel wanted the moment of the vernal equinox, this being the two moments in a year when the sun is exactly above the Equator and the day and night are of equal length, to mark the starting point for the reckoning of a single time throughout the world. Although the *Nautical Almanac* adopted this practice from 1828, Herschel's scheme secured little support.

Nevertheless, Herschel's efforts to discipline the collection of data around the globe proved far more productive than his proposals for an astronomical time standard. Along with the network of physical observatories, each equipped with standardized magnetic and meteorological instruments, Herschel oversaw the production of the Admiralty's textbook of standardized philosophical investigation in 1849. Aimed at naval officers, Herschel's *Manual of Scientific Enquiry* included chapters on a range of natural phenomena. He secured entries on botany, geology, astronomy, magnetism, tides, and a host of other subjects from

Britain's premier scientific authorities including Sabine, Charles Darwin (1809–82), William Whewell (1794–1866), and the Astronomer Royal, George Biddell Airy (1801–92). Together, this project was intended to standardize measurements of natural phenomena from around the world. Herschel's own contribution to his volume, on meteorology, was typical of this ambition. He outlined how naval officers should keep regular hours of meteorological observation during their voyages, specifically at 3 a.m., 3 p.m., 9 a.m., and 9 p.m.. They were to employ the 'systematic process' known as 'keeping a meteorological record' in which readings of all meteorological instruments and weather variables were to be noted without reduction. Only by performing this work throughout an entire voyage could, Herschel warned, the standardized data be produced on which depended 'the development of the great laws of this science'.[13] He also used this publication to encourage naval crews and travelling natural philosophers to take observations of southern-hemisphere stars to compare with northern stars and, therefore, bring about a uniform system of comparative measurement to bear on the entire heavens. Herschel's Manual effectively took the principles of number, counting, and standard units, as outlined in his Preliminary Discourse, and applied them to the global investigation of nature.

A Public Servant of Standardization

Herschel's vision of science, in which the collection of numerical data through precision measurements led to the formulation of the physical laws that governed nature, depended on the creation and acceptance of unifying standards. But nineteenth-century standardization was not just about manufacturing objects of philosophical construction: it was also a crucial socio-political concern, securing commerce and disciplining economic activity. Questions over a nation's units of weight, distance, time, volume, and coinage were very much governmental affairs, involving the regulation of socio-economic life; but in mid-nineteenth-century Britain, it was to scientific authorities that politicians and statesmen looked for unifying standards. As demonstrated through his part in the successful lobbying of Parliament for a magnetic survey, few in early Victorian Britain rivalled Herschel's eminence as a trusted and influential scientific advisor to the government. Between 1838 and 1855, he served in a series of positions in which he guided Parliament and various public institutions over questions of standards.

Almost immediately on returning from the Cape, Herschel joined an ongoing royal commission considering Britain's weights and measures. Back in 1824, a parliamentary commission had found over 230 provincial systems of weights and measures in use throughout the British Isles. In response, Parliament had passed the Weights and Measures Act in an attempt to unify regional variations in units. The result of this was the legalization of the national yard of 36 inches, based on Henry Kater's (1777–1835) measures of a pendulum beating a second at London.[14] The national set of imperial weights and measures, from which all other physical measures were copied, were subsequently stored at Parliament, in the Palace of Westminster, until these were lost in the devastating fire that destroyed the legislative building in 1834. In May 1838, the Chancellor of the Exchequer instructed the Astronomer Royal, Airy, to lead a commission to investigate and recommend how best to replace the nation's weights and measures. Several influential scientific authorities collaborated with Airy on this inquiry, including Francis Baily (1774–1844), Davies Gilbert (1767–1839), George Peacock (1791–1858), and Richard Sheepshanks (1794–1855). At the time of his appointment, Herschel was still overseas, but on returning to England he wasted little time in joining the commission. Over the next three years, Airy's commissioners examined the standard yards and weights that had been recovered from the debris of the Palace of Westminster, but these were so badly damaged as to be useless in determining how long the legal yard had been or how heavy the pound or ounce were before 1834. The commission considered taking some natural reference on which to base the new yard, 'such as the length of a degree of meridian on the earth's surface in an assigned latitude, or the length of the pendulum vibrating seconds in a specified place'. However, they eventually resolved that it was more practical to calculate a new set of measures based on examinations of various weights and measures from around the Britain Isles, including those copied from the corrupted parliamentary standards. Reporting in 1841, Airy's commission provided specifications on a new system of weights and measures, with an imperial yard embodied by a length of metal and embedded in 'the masonry of some public building'.[15] There would be four sets, including yards, ounces, and pounds, which would be compared with each other once every ten years under the authority of a Warden of the Parliamentary Standards. By 1845, Sheepshanks had produced a new measure for the imperial yard, which became the legal standard in 1855.

Along with questions of pounds and yards, Herschel shared in the responsibility for the national time standard, serving on the Board of Visitors for the Royal Observatory at Greenwich from 1843. Drawing its members from the Royal Society, Astronomical Society, and the universities of Oxford and Cambridge, this committee reported to the Admiralty on the state of the observatory, surveyed the institution's instruments, and instructed the Astronomer Royal in his duties. Meeting on the first Saturday of June and, from 1837, the second Saturday of January, the Visitors were tasked with maintaining what was essentially the timekeeper of the nation.[16] Until the 1820s, there was no national standard of time, with different towns and regions taking local measures using the sun. But after 1825, with the growth of Britain's railways, came calls for a single national standard time, taken astronomically at the Royal Observatory, that could be used to regulate the nation's new railway network. By the 1840s, the Post Office and private railway companies were operating on Greenwich time and, from 1851, the observatory transmitted regular time signals through electric telegraph cables, making the rapid communication of the standard a reality. Eventually this time-distribution system would encircle the world thanks to Britain's celebrated, if often technically troublesome, network of submarine cables.[17] Though Herschel was not tasked with performing the measurements on which this standard depended he was, as a Visitor, responsible for ensuring the observatory functioned efficiently and fulfilled its unifying obligations. In this role, Herschel was influential in persuading the government to reject the South Eastern Railway Company's proposals to build a railway line through Greenwich Park. Between 1845 and 1846 this company lobbied Parliament for permission to establish a connection near the observatory but, at a meeting of the Visitors, Herschel warned that passing trains would cause tremors, disrupting the reflecting mercury troughs on which precise astronomical observations depended. When the

Admiralty subsequently refused to allow the railway link, this represented a considerable coup for those eager to preserve the accuracy of Britain's national standard of time.[18]

The most significant of Herschel's public duties, however, was his work as the Master of the Mint between 1850 and 1855. In this capacity, a position that Isaac Newton had once occupied and constituting Herschel's only period of paid full-time employment, he faced the challenge of reforming the institution charged with maintaining the coinage and bullion standards on which the commerce of the British Empire relied. His appointment coincided with a period of upheaval at the Mint, with his predecessor, Richard Sheil (1791–1851), having emphasized to the Treasury the urgency of institutional reforms. In 1849, a parliamentary commission had recommended dispensing with the services of the Mint's moneyers and having the melting of bullion performed by a salaried officer, while ending the practice of contracting out coin production to private businesses. Sheil, however, thought that a contract system was cheapest, ensuring competitive prices. He did not necessarily favour such contracts but did not want the option prohibited. Sheil also wanted several respected individuals appointed assayers to the Mint who could be called on to 'make, within separate laboratories, such independent assays' as the Master might require. On replacing Sheil, Herschel concurred with these economizing recommendations, as well as accepting a reduced annual salary of £1,500, this being £500 below what was customary for the position. In March 1851, Herschel learnt that the Bank of England was having gold refined privately at an establishment in London at a far lower cost than that which the Mint could provide. He therefore resolved to rely 'on the public market for supplies of fine gold' instead of maintaining the Mint's own private, but expensive, refinery.[19] Along with this cost-saving measure, Herschel endeavoured to 'improve the accuracy of the coinage' by purchasing new 'weighing machines on a similar principle to those introduced into the Bank of England'.[20]

Herschel continued to reform the Mint's practices for maintaining gold and silver coin standards in reference to the recent 'progress of the art of assaying'. The legal standard for minted gold specified that up to 12 out of 5,760 grains per troy pound could be composed of silver or copper to ensure the durability of coin. This allowance, however, was often overcompensated for, such as in 1830, when gold sovereigns minted contained 1½ pennyweight of silver per pound in excess of the legal standard. The problem with this was that, providing they had the industrial means, it was profitable for a private individual to melt down the coin and sell off the gold and silver separately as bullion. This had happened during the 1820s, causing a critical shortage of legal tender. Herschel found that much coin in circulation did not meet the legal standard and was forced to embark on a gradual process of re-coinage.[21]

In addition to these extensive reforms to maintain the nation's currency standards, Herschel oversaw a huge increase in the circulation of Britain's coinage. In January 1854, he reported to the Treasury on the expansion of gold monies coined at the Mint in recent years, with less than half a million ounces of sovereigns minted in 1850, rising to over two million issued in 1852 and nearly three million in 1853.[22] There were similar increases for the value of silver and copper coinage minted under his direction. All of this industry demanded exacting standards of precision measurement and careful control over the quality and quantity of bullion employed. In particular, Herschel complained that much gold arriving at the Mint was not of a standard fit for coinage and required refining. It was not, he argued, the Mint's function to act as a refinery, and he subsequently drew up and implemented a new regime for accepting bullion deliveries at the Mint, specifying 'the standard value at which the Master of the Mint is willing to receive it'.[23]

From 1851, Herschel's work at the Mint was made more difficult thanks to the discovery of immense quantities of gold in the British colonies of New South Wales and Victoria. With Australian gold flooding international markets, inflation in the territories surrounding the gold fields out of control, and British coin in high demand for the purchase of unminted ore, the Mint faced a considerable challenge. Herschel not only had to manage a rapid escalation in coin production but also had to secure the integrity of the gold standard on which sterling drew its credibility and oversee the establishment of proposed branches of the Mint in Australia. Following petitions from the legislative councils of New South Wales, Victoria, and South Australia for the founding of minting facilities at Sydney, Melbourne, and Adelaide, Herschel drew up instructions to ensure that colonial mints would produce coinage to the precise standards of the Royal Mint in England. He specified that the dies should be furnished from the Royal Mint itself, that any colonial mint maintain an exact precision in terms of fineness and weight equal to those attained in England, that coins be sent to the Master regularly for testing and assaying, and that he maintain the authority to dissolve any new mint should it fail to ensure identical standards to those achieved in London.[24] The governors of New South Wales and Victoria accepted Herschel's demands, establishing mints at Sydney in 1855 and Melbourne in 1872.

In London, Herschel also provided evidence before a parliamentary select committee tasked with examining options for introducing a system of decimalized coinage throughout the British Isles. In 1853, the committee reported, recommending the adoption of a new standard pound sterling divided respectively into tenth, hundredth, and thousandth parts or, rather, 'florins', 'cents', and 'mils'. Herschel's evidence from the Mint featured centrally within this proposed reform, suggesting that an outright replacement of the nation's copper coinage would be impractical, given the estimated circulation of over 270,000,000 pieces at a weight of some 5,000 tonnes. He preferred the raising in value of older larger penny pieces and the lowering in value of smaller ones. While he thought a new decimalized system preferable to the existing regime of shillings, comprised of twelve pennies, and pennies divisible into quarters, he feared that the lower classes would resist too rapid a change.[25] For all that Herschel was confident that a new decimalized standard of coinage would save clerks and accountants time and labour in calculation, he recognized the limits of such standardizing efforts to reorganize society.

Predictably, Herschel's recommendations aroused considerable hostility as a wave of pamphleteers rejected the committee's proposed system. One critic, taking particular aim at the Master of the Mint, lamented that instead of the project being considered 'by men practically acquainted with the wants and feelings of the mass of the people', it had been examined 'by certain mathematicians from Cambridge, who have proposed an abstruse mathematical system'.[26] Herschel, along with his fellow Cambridge-trained mathematicians, Airy and Augustus De Morgan (1806–71), might have boasted impressive scientific credentials, but such reputation was not the single priority for the formation of standardizing regimes with such direct social impact. Parliament took the committee's proposals no further: Britain would not decimalize its currency until 1971. By 1854, Herschel was on the verge of a mental breakdown and, in 1855, he resigned from the Mint. Thoroughly exhausted from his experience, he never again possessed quite the same industriousness that had characterized his work before 1850. Yet he would remain active within national discussions over standards until his death in 1871.

Standards Celestial and Musical

During the 1860s, the question of Britain's national standards for weights and measures became increasingly politicized. In 1860, Richard Cobden (1804–65) negotiated the signing of a free trade treaty between Britain and France, with French markets opened up to British industrial products in exchange for huge tax reductions on French wines and brandy imported into Britain. Liberal-minded reformers like Cobden and William Ewart (1798–1869) were confident that economic integration would ensure international peace and were keen to encourage greater alignment between Britain and its Continental partners, including the units of measurement that regulated all business and commercial activity. In 1863, Ewart introduced a bill into Parliament, proposing Britain and its empire adopt the French metric system and replace the imperial yard with the metre. The metric system entailed a connected collection of weights and measures, all derived from the metre, with the gram being the weight of a cubic centimetre of water. Despite a heated debate in the House of Commons, MPs voted 110 to 75 in favour of Ewart's motion.

This radical move stirred Herschel from his post-Mint retirement, as he launched a fervent campaign against the metre. During the 1840s, he had been sympathetic towards the principles of metrication; but with the new legal yard established in 1855, Herschel was committed to a standard he felt well-supported by a great deal of experimental labour. In October 1863, he lectured to Leeds Astronomical Society on the failings of the metric system before circulating a pamphlet laying out these arguments among MPs down at Westminster.[27] Herschel here drew attention to Parliament's recent legislation to abolish Britain's existing weights and measures, which had passed its second reading and was, effectively, a political commitment. Yet this was a crucial social question that demanded greater public attention than the few notices that had appeared in *The Times*. Herschel outlined the character of a suitable national standard, before comparing the yard with the metre, and then discussing which was best and the system of numerical multiplication and subdivision most appropriate for such a unit. In the past, ancient societies had selected the physical features of heroes or rulers as unifying measures, such as the height of Goliath the Philistine, or cereals, with the English inch supposedly three grains of barley. Another course was to employ the dimensions of a typical man, but as the average Belgian had been 5 feet 7 inches in 1817, compared to an average French conscript of 5 feet 4 inches, and an average Lancashire non-manufacturing labourer of 5 feet 10¾ inches, such a standard would clearly not do.[28] Similarly, were such a human or agricultural unit selected but then lost through some calamity or a gradual change in climate, the measure would be lost or lose all relevance. Instead, Herschel called for '*a universal* standard', this being some object, natural or artificial, that was imperishable in nature and would survive the decaying influence of time. It had to be a measure universally true in all places and at all times and had to have a 'claim to *general* acceptation as of common interest to all mankind, or at least to all the civilized portion of it'.[29] Above all, an ideal standard had to be drawn from the standardization evident in nature: in other words, the units of Creation and its divine architect. Herschel felt that, in this respect, the metre failed completely.

The only solution that met Herschel's requirements for an authoritative standard was one based on a natural object and, he contended, the one object in nature that most perfectly combined the demands of a such a standard was the globe itself. Its unchanging dimensions provided two naturally defined lengths free from 'geological revolutions and catastrophes', these being the Earth's polar axis and its equatorial circumference.[30] Of these, given that the equator was not strictly circular, Herschel favoured basing Britain's measure of length on the distance of the axis passing directly through the centre of the Earth from the north to the south pole. There were, he claimed, three theoretically superior natural measures: the velocity of light, the length of a light ray's undulation, or the length of a pendulum beating seconds under specific conditions. While the two light-based units seemed to Herschel to be 'purely visionary', a pendulum-based unit was possible, this being

> the space fallen through by a heavy body on the same place by the earth's attraction in a second of time. The modulus so obtained is therefore a measure of the earth's total attractive power . . . as that derived from the length of its diameter is of its total bulk, and equally unalterable and universal.[31]

The English yard of 1824 had been based on such measures, taken at sea-level in London, but this had been found to be erroneous. Herschel thought recovering this unit, lost in the fire of 1834, too geographically variable, being subject to differences in the Earth's density. Nevertheless, this unit had the advantage of linking space and time. However, Herschel noted that in 1798 the French commissioners responsible for the metric system's creation rejected such a connection because a second represented an 86,400th part of a day and could not be decimalized. They had instead based the metre on calculation of a ten-millionth part of a quadrant of the Earth, taken along a meridian from the north pole to the equator. Yet this was, Herschel maintained, a poor standard. As a measure of the meridian running through Dunkirk and Formentera, it was neither universal nor exact, as Airy's calculations of the Earth's various meridian arcs had shown in 1830. Recent surveys showed that the distance of the Earth's arc to be 41,711,019.2 feet as it passed through Russia, compared to 41,697,496.4 feet as it passed through France.[32] It would be irrational, thought Herschel, to swap Britain's arbitrary yard for such a subjective metre, especially as Britain's weights and measures regulated trade throughout the British Empire, North America, and the Russian Empire and constituted the most extensive 'commercial relations' in the world. It would, he reasoned, make more sense for France to adopt the imperial system.[33]

Herschel personally favoured a new unit altogether, based on the measure of Earth's polar axis. Given that the distance from the north to south pole, running directly through the Earth, was reckoned to be 500,497,056 imperial inches – just 2,944 shy of 500,500,000 – he recommended increasing the length of the imperial yard by a 1/

1000th part. This change would be unfelt commercially or by engineers and architects. A 'geometrical inch' would be an exact 500,000,000th part of the earth's polar axis, and a rod of 25 of these inches, 1/10,000,000th part of the polar semi-axis, would give a 'geometrical cubit'.[34] This unit had astronomical credentials, as Herschel claimed that there were 'several links which connect the *British standard* inch with the distances of the fixed stars, and of the sort of intermediate units we have to deal with in the inquiry'.[35] In 1864, Parliament voted again and revised its decision to make the metric system compulsory, opting instead for a permissive act, but in 1869 the Commons again moved to establish the French measures as Britain's legal standards. Again, Herschel was appalled, condemning the 'too hasty preference for the apparently more scientifically, and certainly more, symmetrically, constructed system of our continental neighbours'.[36] He reiterated the claims of his 'geometrical inch', derived from the Earth's polar axis but, as in 1864, found little enthusiasm for the standard.

If Herschel's attempts to reform the inch roused little political support, his quixotic plan to reorganize the nation's musical standards according to mathematical principles was similarly unsuccessful. If the failure of his reformed inch was largely due to a lack of desire among Britain's politicians to reorganize the existing system of measurement, then the lack of support for Herschel's reformed pitch was largely down to his inability to reconcile mathematical theory with the aesthetic demands of the nation's musicians. Throughout the early nineteenth century, there was broad consensus among European musical communities that the pitch at which orchestras were playing music was rising. In the mid-1830s, an A-above-middle C in Paris was generally sounded between 435 and 443 vibrations per second, but by the 1850s, most institutions played at a higher frequency. The Paris Conservatoire's A had increased from 436 to 446 between 1847 and 1856, while the Lille Opéra was at 451 by 1854.[37] In 1858, the French state attempted to resolve this musical escalation, with Emperor Napoléon III (1808–73) sanctioning a commission to introduce musical uniformity throughout the country. A year later, this inquiry recommended A435 as the national unit of pitch, which Napoléon's government passed into law as the *diapason normal*.

British musical and scientific audiences responded with a similar scheme for eradicating pitch disparities. Several months after the announcement of the French measure, in mid-1859, the Society of Arts established a committee of musical and scientific authorities to agree on a British standard, but this enterprise rapidly became controversial. While some favoured following France's standard of A435, which would equate to a C of 522, others promoted a higher pitch altogether, preferring the high 520s or low 530s: generally, a higher frequency was thought to be more pleasant on the ear than a lower one. During its proceedings, the Society of Arts's pitch committee received a letter from Herschel, read at a meeting in July 1859, making a case that the new standard should be based on mathematical principles, rather than musical aesthetics.

Though musicians favoured higher, brighter frequencies, Herschel demanded the appointment of C512 as the nation's musical standard. This was based on the principle that if a string vibrating the lowest possible C oscillated once per second, each following C in musical scale would increase to the power of two: C512 would represent the ninth C in this succession, or C-above-middle C. Herschel argued that C512 was a scientific, natural standard which,

> being the ninth Octave of a fundamental note corresponding to one vibration per second, has a claim to universal reception on the score of intrinsic simplicity, convenience of memory, and reference to a natural unit, so strong that I am amazed at the French not having been the foremost to recognize and adopt it; when it is remembered that their boasted unit length, the metre, is based on the subdivision of a natural unit of space, just as a second (a universally used aliquot of the day) is of time: the one on the linear dimensions, the other on the time of rotation of the Earth.[38]

This was a grand standardizing vision: Herschel wanted the units of length, time, and music to be united in reference to universal phenomena. What was required, he contended, was a musical pitch which took nature as a standard, and this involved basing any measure on the vibration of a string. He confessed himself 'to be more French than the French themselves' in his enthusiasm for C512, which would make musical pitch 'part and parcel of a complete natural metrical system which would recommend itself to all nations on its own merits'.[39]

Although considering Herschel's arguments seriously, the committee thought such a measure impractical and aesthetically poor for listeners. Instead, they favoured a higher, brighter sounding unit. On 1 June 1860, the pitch committee published its report, proposing C528 as the national standard, and held a public meeting four days later at the Society of Arts to discuss its implementation. Clearly disappointed with the rejection of his mathematical standard, Herschel himself turned up to this gathering and made a final speech in favour of C512 but, with the audience consisting largely of musicians, music theorists, composers, and patrons of the arts, the meeting's participants rejected his amendment, voting against any introduction of a mathematical standard.[40]

Herschel was not only eager to standardize the pitch to which musicians tuned their instruments, he also envisaged an entirely reformed system of musical measurement. Eight years after his failure to establish C512 as the national British unit, in 1868, Herschel proposed a new scale of musical notes based on ratios of vibration and expressed through an octave divided into 1,000 equal intervals. Originally published in the *Quarterly Journal of Science*, he argued that the existing harmonic system of fractions, with notes divided into octaves, fifths, thirds, and sevenths of a fundamental note, was irrational. A mathematically superior harmonic standard, he explained would be for the fundamental note, C for instance, to be set at 0, and its octave to be 1,000. That would make D 170 parts, E 322 parts (rather than a third), F 415, G 585 (instead of a fifth), A 737, and B 907.[41] As with his proposals for a new standard pitch, Britain's musical communities were understandably uninterested in considering such a harmonic reorganization. For all that Herschel held sway with scientific and parliamentary audiences, extending his standardizing aspirations to broader society proved unsuccessful.

Herschel's interventions into questions of metrication, musical pitch, and harmonic division were futile. He failed to exert influence over standardizing projects that were ostensibly social, political, economic, or artistic concerns. When he spoke, there were audiences, such as MPs and Society of Arts committee members, that wanted to know what he thought. But his claims of the astronomical precision of a geometrical inch or the mathematical credentials of a musical C of 512 vibrations failed to build enough consensus to become universal, or even national, standards. For all that Herschel idealized standards within scientific practice and extorted their importance within the measure of nature, his views represented just one position within incredibly complex processes of standardization. Herschel's universe may have been a divinely created place of order, but social and political audiences proved far more troublesome to discipline according to scientific principles and precise measurement.

Conclusion: Standardizing Nature

Herschel died in 1871. In the same year, James Clerk Maxwell (1831–79) accepted Cambridge's recently inaugurated chair in physics and began working to establish the University's new physical laboratory, subsequently known as the 'Cavendish'. Completed in 1873, this new institution was the latest in a series of physical laboratories built throughout Britain during the mid-nineteenth century, including at Glasgow in the late 1840s, Oxford and University College London in 1866, and King's College London in 1868. Stimulating this expansion in laboratory culture was the production of electrical standards for Britain's growing telegraphy network. Although promising rapid communication, the cables that conveyed the electrical signals on which this imperial system depended had a tendency to break or fail, while locating faults was difficult without accurate units of measurement for electrical resistance. Laboratories like the Cavendish were very much a response to this problem: they were idealized sites of measurement, fashioned for the production of industrial standards.

In a sense, these laboratories were the built manifestations of Herschel's vision of science as the practice of counting and precision measurement. In 1830, Herschel had famously declared the universe to consist of countable and ordered atoms, each bearing the character 'of a *manufactured article*' and a creating power.[42] For Maxwell, the challenge of placing a physical laboratory within the Liberal Anglican traditions of the University of Cambridge was considerable, given the industrial connotations of such an establishment and its intended focus on ensuring ever more precise standard units of natural phenomena. In his address for the Cavendish's opening, he declared that in 'experimental researches, strictly so called, the ultimate object is to measure something' and that a laboratory risked being 'out of place in the

University, and ought rather to be classed with the other great work-shops of our country'.[43] Such a reduction of philosophical inquiry to the pursuit of number, measure, and weight threatened to result in a theologically materialistic understanding of the universe: a laboratory of this kind would be little more than a factory. It was, however, to Herschel that Maxwell looked for a solution. For Herschel, measurement revealed the divine structure of nature through its evident order.[44] To measure atoms was to scrutinize the molecules out of which nature was created and, 'as Sir John Herschel has well said', each atom bore 'the essential character of a manufactured article'.[45] Herschel's philosophical programme for ascertaining laws of nature through measurement and counting and the use of universal standards came to increasingly dominate nineteenth-century scientific culture, but this fit firmly within his own portrayal of a divinely created universe.

Notes

1 John Herschel, *A Preliminary Discourse on the Study of Natural Philosophy* (London: Longman, Rees, Orme, Brown, and Green, 1830), 124.

2 Herschel, *A Preliminary Discourse*, 125.

3 Herschel, *A Preliminary Discourse*, 126.

4 Herschel, *A Preliminary Discourse*, 123.

5 Herschel, *A Preliminary Discourse*, 122.

6 Stephen Case, '"Land-Marks of the Universe": John Herschel against the Background of Positional Astronomy', *Annals of Science*, 72.4 (2015): 417–34, 422; Stephen Case, *Making Stars Physical: The Astronomy of Sir John Herschel* (Pittsburgh: University of Pittsburgh Press, 2018), 32–54.

7 An officer in the Royal Artillery, Sabine had been on several naval expeditions during the 1810s and 1820s, recording natural phenomena, including those magnetic. In 1828, he became one of three scientific advisors to the Admiralty and rose rapidly within the Royal Society, eventually becoming its president in 1861.

8 John Cawood, 'The Magnetic Crusade: Science and Politics in Early Victorian Britain', *Isis*, 70.4 (December 1979): 493–518, 507–10.

9 Anonymous, 'Fifteenth Meeting of the British Association for the Advancement of Science. Cambridge, 18th June 1845', *Calcutta Journal of Natural History: And Miscellany of the Arts and Sciences in India*, Vol. VII (Calcutta: Bishop's College Press, 1847), 81–142, 93.

10 For a discussion of these tensions between those who emphasized expeditionary, and those valuing observatory, magnetic research, see

Christopher Carter, *Magnetic Fever: Global Imperialism and Empiricism in the Nineteenth Century* (Philadelphia: American Philosophical Society, 2009), 46.

11 Herschel quoted in Christopher Carter, 'Herschel, Humboldt and Imperial Science', in Raymond Erickson, Mauricio A. Font, and Brian Schwartz (eds.), *Alexander von Humboldt. From the Americas to the Cosmos: An International Disciplinary Conference, October 14–16, 2004* (New York: The City University of New York, Bildner Center for Western Hemisphere Studies, 2004), 509–18, 515.

12 Mary Somerville, *On the Connection of the Physical Sciences* (New York: Harper & Brothers, 1846), 81.

13 John Herschel, 'Meteorology', in John F. W. Herschel (ed.), *Manual of Scientific Enquiry; Prepared for the Use of Her Majesty's Navy: And Adapted for Travellers in General* (London: John Murray, 1849), 268–322, 269.

14 Simon Schaffer, 'Metrology, Metrication, and Victorian Values', in Bernard Lightman (ed.), *Victorian Science in Context* (Chicago: University of Chicago Press, 1997), 438–74, 443.

15 Report of the Commissioners Appointed to Consider the Steps to Be Taken for Restoration of the Standards of Weight & Measure (London: Her Majesty's Stationary Office, 1841), 5–7.

16 S. Laurie, 'The Board of Visitors of the Royal Observatory – II: 1830–1965', *Quarterly Journal of the Royal Astronomical Society*, 8.4 (December 1967): 334–53, 337.

17 Derek Howse, *Greenwich Time and the Longitude* (London: Philip Wilson, 1997), 89–96.

18 Edward J. Gillin, 'Tremoring Transits: Railways, the Royal Observatory and the Capitalist Challenge to Victorian Astronomical Science', *British Journal for the History of Science*, 53.1 (March 2020): 1–24, 14.

19 *Mint. Copy of any reports to show the changes that have been made in the management of the Royal Mint, and its present state*, House of Commons Papers, No. 76 (1852), 3–4.

20 *Mint. Copy of any reports*, 6.

21 *Mint. Copy of any reports*, 10.

22 *Mint. Account 'of all Gold, Silver, and Copper monies of the realm, coined at the Mint, from the 1st day of January 1848 to 31st day of December 1853'*, House of Commons Papers, (1854), 2.

23 *Mint. Copy of any reports*, House of Commons Papers, No. 76 (1852), 55.

24 Australian Mints. Copy of Treasury Minute, Dated March 22, 1853. Parliamentary Paper (London: George E. Evans and William Spottiswoode, 1853), 2–3.

25 *Report from the Select Committee on Decimal Coinage; together with the proceedings of the committee*, House of Commons Papers (1853), 45–62.

26 John Edward Grey, *Decimal Coinage: What It Ought and What It Ought Not to Be. By One of the Million* (London; James Ridgway, 1854), 4.

27 Schaffer, 'Metrology, Metrication, and Victorian Values', 448.

28 John F. W. Herschel, *Familiar Lectures on Scientific Subjects* (New York: George Routledge & Sons, 1869), 424–25.

29 Herschel, *Familiar Lectures*, 426.

30 Herschel, *Familiar Lectures*, 427.

31 Herschel, *Familiar Lectures*, 429.

32 Herschel, *Familiar Lectures*, 433 and 439–41.

33 Herschel, *Familiar Lectures*, 445.

34 Herschel, *Familiar Lectures*, 447–50 and 191–92.

35 Herschel, *Familiar Lectures*, 181.

36 Herschel, *Familiar Lectures*, 178.

37 Bruce Haynes, *History of Performing Pitch: The Story of 'A'* (Lanham: The Scarecrow Press, 2002), 346–47.

38 John Herschel, 'Uniform Musical Pitch', *Leeds Mercury* (Leeds, England) 2 August 1859; Issue 6985.

39 Herschel, 'Uniform Musical Pitch'.

40 Edward Gillin and Fanny Gribenski, 'The Politics of Musical Standardization in Nineteenth-Century France and Britain', *Past and Present*, no. 251 (May 2021): 153–87; for a history of pitch standardization, see, Fanny Gribenski, *Tuning the World: Acoustics, Aesthetics, Industry, and Global Politics* (Chicago: Chicago University Press, 2023).

41 John Herschel, *On Musical Scales* (London: W. Clowes and Sons, 1868), 4.

42 Herschel, *A Preliminary Discourse*, 38.

43 Maxwell quoted in, J. G. Crowther, *The Cavendish Laboratory, 1874–1974* (London: MacMillan, 1974), 38.

44 Schaffer, 'Metrology, Metrication, and Victorian Values', 446.

45 Maxwell quoted in, W. D. Niven (ed.), *The Scientific Papers of James Clerk Maxwell*, vol. 2 (Cambridge: Cambridge University Press, 1890), 376.

10

John Herschel and Politics

John Herschel was a political radical in the utilitarian mould but no social leveller. Indeed, he was highly ambivalent over the notion of 'democracy' and feared it could unleash the tyranny of the masses. Rather, like his one-time close friend Charles Babbage, he was a rational materialist reformer with little sympathy for old state props such as the Church of England and the prevailing curriculum of the elite Anglican universities. This periodically put him at logger heads with contemporary Cambridge and Oxford taught men. (Herschel had graduated from Cambridge as senior wrangler in 1813 and Babbage was awarded a Cambridge degree with no examination in 1814.)

Herschel, like Babbage and many others during this time, saw intelligence, or the way one came to objective knowledge, in mechanistic terms described by symbolic algebra.[1] In other words, intelligence was a process – like algebra – that in the abstract was incontrovertible and excluded any metaphysical unknowns such as divine origins.[2] This was faith in human, not divine, reason. Privately, Herschel believed there was no room for God and heavenly intervention in such a perspective. Likewise, politics and its accompanying institutions should be organised and structured along such rationalist lines – what Herschel referred to as the 'greater logic' in his *Preliminary Discourse on the Study of Natural Philosophy* (1830).[3] In this sense, Herschel was on the same page as the recent French Revolution – albeit no democrat.

However, unlike Babbage, Herschel evaded confrontation rather than provoking it. It was no coincidence that soon after the failure of his election as the next president of the Royal Society in 1830 and his role in the Great Reform Act of 1832, he fled to South Africa. (Although, to be sure, this retreat was a typical, highly successful, and characteristic ploy by Herschel to use colonial experience to bolster public repute in the metropole.)[4] Thereafter, partly through coaching by his new pious wife Margaret Brodie Herschel, he worked tirelessly, whenever possible, to avoid controversy. This meant later distancing himself from contentious scientific figures such as his onetime friend Babbage, Charles Darwin, and anyone who sought to publicly rethink human nature. This chapter will set out to explore Herschel's politics, how it sat in the prevailing historical context, and his notion of knowledge. All this, as we shall see, was informed by his idea of intelligence defined by his perceived method of arriving at objective knowledge.

Mathematical Reform

The first indication of Herschel's politics appeared during his early days at Cambridge. The Mathematical Tripos dominated education when Herschel arrived at St John's in 1809. But what type of mathematics should be taught? Surely mathematics was politically neutral? As we shall see, however, the introduction of French algebraic analysis, such as that advocated by the Analytical Society (1812–14), was disturbing to several key Cambridge dons.[5] As the key founders of this society, Herschel and Babbage came to represent the antithesis of the university's Anglican mathematicians.

Herschel's early radicalism has tended to get lost in the search for precursors to the current emphasis upon machine intelligence in the work of Babbage. To begin with, at least, Herschel was equally enthusiastic about both machine intelligence and mathematical reform. By the early 1820s, his contemporary, the Trinity student and ultra-High Anglican Hugh James Rose (1795–1838), despised proponents of such a mechanistic philosophy, which, Rose felt, led to 'scepticism as to the existence of a first Cause'. Moreover, Rose fumed:

> I hate both alike as leading to the cultivation of the sensible part of man instead of his spiritual & teaching people to occupy themselves with the theory of differences & the knowledge of formulae, which if a man knew all about them that was or can be known, leave him just where he was

not wiser but fuller – not with a more occupied understanding: leaving much instead the same sort of thing as Babbage's calculating machine – Indeed I believe that the approximation of the operations of the mind & of machinery is a very fit occupation for the Mechanic school.[6]

A few years later, in 1828, Herschel felt he had to defend his colleague in mathematical reform against media accounts regarding the costs involved in Babbage's calculating engine project and unfounded suggestions that Babbage was pocketing the money for himself. Herschel clearly saw this, too, as an attack upon their belief in symbolic algebraic reasoning and an implication that his friend was imprudent.[7]

For many at Cambridge, the emerging mechanical world expounded by people like Herschel took society further and further from its divine, natural, simple, rural, and hierarchical state; the laws of nature, Herschel's Anglican opponents argued, could not be found via an abstract logic or simple sensory science (inductive observations). For Herschel, like Babbage, reasoning with algebraic symbols was mechanistic and debates over imaginary numbers were completely redundant in trying to refute such a system.[8] Before George Peacock (1791–1858) made the case that symbolic algebra was not confined to arithmetic and Augustus De Morgan (1806–71) and, especially, George Boole (1815–64) applied it overtly to logic, Herschel and Babbage had made the case in less manifest terms.[9] By contrast, the prevailing Anglican establishment claimed the further one got from the human scarred landscape – be it physical or intellectual – the closer one got to God; an expanding knowledge of the outside world depended on a growing knowledge of the internal consciousness. This debate had major political implications on the way lives were ordered.[10] Emotion was the link between the inside and outside world with the act of introspection achieved via the imagination.[11] By contrast, there was no room for emotion or some divine entity in Herschel's notion of mechanistic intelligence.

This belief in the role of divine ideas as central to knowledge lay at the heart of the views of Herschel's contemporary at Cambridge – the Trinity-educated William Whewell. But the division between their views had not always been so clear; for example, on arriving at Trinity, the indiscriminately enthusiastic Whewell was initially attracted to the Cambridge Analytical Society. Set up by Herschel and

Babbage, the Society was principally created to challenge what they considered to be the dilapidated intellectual climate of the university.[12] The aim of the Society was to import the most powerful techniques of mathematical analysis and replace the Newtonian fluxional notation with the d's of Leibnizian differentials. Its inspiration derived from French mathematicians such as Pierre-Simon Laplace (1749–1827), Sylvestre Lacroix (1765–1843), and especially the powerful analytical machine of the algebraic calculus of Joseph-Louis Lagrange (1736–1813). The Analytical Society believed Britain should be importing and adopting these predominantly French developments (see Chapter 2).[13]

During his time at Cambridge, Herschel veered toward what the university hierarchy would have considered politically radical views along the lines of the recent French Revolution. It is quite wrong to imply, as some historians do, that Herschel, Babbage, Whewell, and Richard Jones (1790–1855) formed a coherent intellectual group unified in their views toward science, politics, and reform. As we shall see, the situation was very different, with important and sometimes acrimonious divides emerging quite early between their positions.[14] Herschel and Babbage's vision of the world likely made Whewell and Jones feel quite queasy during their alleged breakfast exchanges – particularly in their final undergraduate days. By this time, they would have become aware of the radical implications of Herschel and Babbage's views and the consequences of their analytical approach to thought. Jones and Whewell had by then become hostile sceptics of such views. The one thing on which they were all united, however, was the importance of natural philosophy and the collection of trustworthy observations – be they historical, economic, social, astronomical, or philological. The differences between them arose from the way those data were to be interpreted and induction was understood.

Herschel and Babbage were moving, via symbolic algebraic analysis, to a mechanistic rational logic. The brain, if correctly disciplined, was perfectly formed to mechanically process and interpret empirical data. Both thus emphasised experience and a version of John Locke's theory of a tabula rasa with observations being sculpted by a mechanistic logic. Operating on algebraic symbols, they believed, put the naked workings of the brain on show. Thought could be rendered visible and available to empirical scrutiny and therefore open to the possibility of being built as a machine; it was no longer a mystical and sacred sphere as Whewell's ilk maintained. This was a quest to which Babbage, initially aided by Herschel, would devote the rest of his life.[15] Here Babbage and Herschel were also drawing on the semiotic philosophy of the Frenchman Étienne Bonnot de Condillac (1714–80) and his impact on Lagrange, Laplace, and Lacroix. Condillac wrote in 1780 that 'the technique of reasoning is thus the same in all the sciences. As, in mathematics, we state the problem by translating it into algebra; so in the other sciences, we state it by translating it into the simplest expression possible'.[16]

However, it is incorrect to say it was the Cambridge Analytical Society that first imported French algebraic analysis into Britain. This mathematical approach was already well established in the British schools based in large military complexes such as Woolwich and among, predominantly, Scottish mathematicians. The military necessities of killing and plotting ballistics did not recognise ideological protests in the same way traditional Anglican universities did.[17] Indeed, Herschel was taught the basics of algebraic analysis by his Scottish home tutor Alexander Rogers, hired by his Hanoverian and deist father, William Herschel. Rogers was patronised by the Edinburgh mathematician and natural philosopher, John Playfair (1748–1819) and at one point offered a position at the Military Academy at Woolwich that he declined in favour of keeping his accountancy job with the London and Edinburgh Shipping Company. Playfair was a staunch critic of the Anglican universities and, no doubt, made an impact upon Herschel via Rogers.[18]

Most Cambridge dons, like their intellectual ancestor Isaac Newton, advocated ancient geometrical techniques that ensured demonstration always applied to something real and revealed the complicated intellectual inductive journey to a theorem. This was discovery not via some mechanical logic but a hard intellectual slog that, not infrequently, moved in unpredictable ways. By contrast, symbolic algebraic analysis would move into the abstract and only return to the original topic once a conclusion was reached. The traditional mathematics taught at Cambridge had developed within an environment designed to produce clergymen for the Church of England; in this sense geometry was as much a national institution as the Church. Samuel Taylor Coleridge (and Whewell), for instance, understood and emphasised this by underlining Euclidian geometry.[19]

Teaching students predominantly French algebraic analysis would, many feared, see Britain's future leaders light the road to a French Revolutionary system of politics founded on abstract thought rather than anything real. This was an argument close to that made by the anti-Jacobin professor of natural philosophy at Edinburgh University, John Robison (1739–1805). Another Edinburgh-based philosopher, William Hamilton (1788–1856), attested that Robison 'preferred the ancient method of studying pure geometry, and even felt a dislike for the Cartesian method of substituting symbols for the operations of the mind, and still more was he disgusted with the substitution of symbols for the very objects of discussion, for lines, surfaces, solids, and their affection'. Herschel, in turn, was disgusted by Robison's work, describing it as 'horrid nonsense', 'abominable', and 'most vile'.[20]

It was not that the conservative Anglican faithful were ignorant of algebraic analysis; it could be a useful tool, but making it central to human understanding made it dangerous and devoid of God. It certainly was not an instrument for the discovery of new knowledge. True reason required and was derived through Anglican divine faith. Bad mathematics was bad theology. During the Napoleonic Wars, for instance, one writer in the *Eclectic Review* (1808) raged:

> The writings of some eminent French mathematicians abound in infidel principles. Our elder men of science, we hope, are for the most part of too sober a cast to be injured by these principles; but we tremble for the fate of the young, and should rejoice to think it probable that these remarks might prevent one such student from abandoning the aids of religion, and venturing himself on the stormy ocean of life, without rudder, without compass, and without hope.[21]

Whewell and his High Anglican allies retained a belief that human understanding had become corrupted after the Fall and that the only way to obtain a glimmer of truth was to access the divine ideas still within the human mind.[22] Our knowledge may have changed, but our damaged intelligence was no better. The objective of education should be preparation for the next, not this, world. By contrast, Herschel wrote in his undated notes on 'Great Physical Theories' that 'the history of science in all its branches ... shews that every great accession to theoretical knowledge has uniformly been followed by a new practice, and by the abandonment of ancient methods as comparatively inefficient and uneconomical'. Objective knowledge, Herschel believed, progressed the closer it came to mechanistic reason. Herschel embraced the new urban-industrial world of the early industrial revolution and saw the need to parallel this in human intellectual development, that is, the material process of reasoning. Intelligence was a mechanical logic – a process all science should eventually meet – not the discovery of divine innate ideas.[23]

Machine Philosophy

In 1810, Herschel visited the Boulton and Watt Soho Foundry with his father. He was ecstatic as they were shown around by his father's good friend, James Watt (1736–1819). Herschel wrote, 'this foundry is famous all over Europe, and as it were the source of every improvement in machinery which has displayed the power and ingenuity of man'. The key to this human design was Watt. His memory, Herschel reflected, was 'stored with every part of elegant, useful, and in many cases profound science, an invention ready in conceiving the most complicated plans of an activity, prompt in their execution, these are even separately very rare, but united they form a prodigy'. It was this analytical mind that reflected the world, not any divine ideas. As Herschel put it in his unpublished philosophical notes: 'There are not two or more systems of Physical truth nor of Logical representation of it.' Human minds all reflected the same universe and could be reduced to the logic of algebraic analysis. To Herschel, Watt's mind was organised exactly like the Soho Foundry; to view the foundry was to see a great mind.[24] He summed it up in his unpublished notes thus: 'Economy is the applying [of] everything to its best use.' Hence, 'Economy of time – This is to time what industry is to power.' In short, speed and thus the efficient application of a division of mental labour was vital: 'Power acting rapidly – it is resultant equality compounded of power and Time. Much Power x Little Time.' Herschel believed that efficiency could be represented as a mental power over time.[25]

For Herschel, new knowledge was obtained once the outside world fitted the abstract rules of symbolic algebraic analysis. This was not necessarily anti-theological; rather, God had departed but had left this form of reason in humans to interpret the unfolding of divine design. In the beginning was not the word but the human faculty of abstraction. As Herschel reflected, 'It is not truth that we have in the abstract – that we must find – but we have the faculty of abstraction to enable us to find it.'[26] Algebraic symbols did not depend on an external referent for their meaning but simply on their initial relationship within the abstract system. This was faith in human logic. Whereas a mechanistic rationality made the observations fit algebraic analysis, geometry waited until the observations formed or recognised a clear idea. In other words, for Herschel it was by the mechanical, engine-like clunking of the human brain alone – without the need for divine innate ideas – that knowledge was discovered.

Herschel made it clear in his essay 'Mathematics' for the *Encyclopaedia Metropolitana* (1830) that, as such, a mathematical approach required 'no assistance from inductive observation, and very little from the evidence of our senses'. It was through the brain's correct algebraic logic that observations could be guided and confirmed.[27] This was a Lockean-inspired point he reiterated in 1837 in the diary he kept at the Cape: 'In the Mass of facts relations exist as Statues exist in Marble – It is the mind which chisels them out and gives them body – by the instrumentality of abstract terms which are the tools and the inward perception of harmony and beauty which guides them.'[28] The mind, through abstract symbolic algebra, was being rendered visible and open to empirical scrutiny in exactly the same way as an idealised manufacturing site – such as his interpretation of the Soho Foundry.

Herschel's religious dissenting views and mechanistic notion of intelligence were later to cause anxiety for his wife, Margaret.[29] The anticlericalism and Unitarian sympathies of his youth tended to be kept private but did manifest in correspondence with his fellow college friend, John William Whittaker (1791–1854). To Whittaker, Herschel wrote that he hoped 'religion, as established by law may never entirely usurp the superiority and control over religion established by nature', representing the material and utilitarian side of William Paley's teachings. Herschel was distinctly hostile to the raging Christian evangelical movement: 'Let them go on, poor devils! Let them convert the Jews, and pour forth the bitter vials of religious controversy ... of religious fanaticism ... so long we are not obliged to stake our happiness and lives against their blindness.' Indeed, William Herschel – despite his own deism – worried about his son's sentiments: 'You say – I cannot help regarding the source of church emolument with an evil eye .... You say the church requires the necessity of keeping up a perpetual system of self-deception or something worse, for the purpose of sup-porting the theological tenets of any particular set of men.'[30] This was simply a practical concern – what other career was there for his son other than the Anglican clergy? A legal one? Herschel tried the latter but hated it.

On his voyage to the Cape, Herschel condemned a recent, popular book entitled *The Sacred History of the World: As Displayed in the Creation and Subsequent Events to the Deluge; Attempted to Be Philosophically Considered, in a Series of Letters to a Son* (1832). This was written by the High Anglican historian Sharon Turner (1768–1847) as an attempt to reconcile religion and science. Herschel described it as 'a vile trash-book'. A far more impressive book than Turner's 'trash' was *The Rationale of Punishment* (1811 and translated from French to English in 1830) by the arch dissenter and utilitarian, Jeremy Bentham (1748–1832). Herschel described Bentham's chapters on 'free competition' and 'rewards for virtue' as displaying 'wonderful acuteness & closeness of argument'. He later reiterated his utilitarian beliefs when he defined 'justice' as 'that quality of actions which – *on the great average* inflicts the least amount of avoidable evil on the parties concerned in them'.[31] Turner's son, to whom his book was addressed, went on to be educated at Trinity, took holy orders, and became a rector. Though, as noted above, Herschel's great dislike for the hierarchical nature of the Anglican Church caused him to disap-point his own father's hopes, he was very impressed by the churches in Cape Town. In contrast with ostentatious London cathedrals, the churches of the Cape had 'few & small pews' and were 'occupied by Chairs & benches, in which all sit indiscriminately, [without] distinc-tions – Everyone should forget Differences & remember only that he is a worm among worms'.[32] Herschel's anti-democratic views clearly dissolved in such places.

Herschel had earlier, at Cambridge, made it clear that an emphasis on tradition and faith in the word of Scripture was not necessary or indeed of any use in the manufacture of objective knowledge. In fact, it was defunct. He wrote, probably during this time: 'The formative and generalising power of the mind connects the dots of collective propositions into the outline or contour of an inductive one. It is the same power which makes *bodies, instances, systems*, theories and insulated perceptions or suggestions given us by repeated acts of attention. It is this which gives us the certainty of truths derived from the differential calculus.'[33] Herschel's emphasis on human logic defined by symbolic algebraic analysis formed a dangerous alliance. Acknowledging the dead end for his reforming ambitions at Cambridge, he escaped to London to pursue his analytical reforms with Babbage in a more conducive environment. This commenced with the establishment of the Astronomical Society of London in 1820 and involvement with more sympathetic metropolitan circles.[34]

The establishment of this new Society was probably the most important development in the reform of the sciences during the first half of the nineteenth century. It quickly challenged the Royal Society as the main institution called upon to offer technical advice to the government. Its originators were, in some ways, the embodiment of Britain's fiscal-military complex, composed of financial men, military officers, and a few young dissenting reform-minded Anglican university products such as Herschel. All emphasised precision measurement and robust inductive empiricism. To be a man of science at this time meant to do astronomy. This made Herschel perfect and of blue blood since his father was Europe's leading astronomer. John Herschel spent much of the 1820s plotting double stars with his fellow Society member, the wealthy and indignant surgeon, James South. During the 1820s two of Herschel's closest friends were South and Babbage – both of whom spearheaded the decline of science argument and sought reform of the Royal Society. However, reform meant far more than just the state of science.

The Great Reform Act

The Great Reform Act of 1832 was a significant moment for reasons beyond simply setting the precedent of expanding the franchise; it challenged medieval voting requirements and assumptions about parliamentary representation along new 'rational' grounds.[35] The impetus of the Boundary Commission, the body set with the task of defining franchise boroughs and the body central to the entire political reform process, should not be reduced to simply a utilitarian thrust spearheaded by Henry Brougham (1778–1868), the Society for the Diffusion of Knowledge, and University College London (although these were clearly important factors).[36] For example, of the twenty-one Boundary Commission members, nine had gone to Oxford and Cambridge – with at least four from Trinity. Many of these were members of the Astronomical Society.

This Society, of which Herschel – as we have seen – was an important founding member, was one of several new scientific societies established during this period as a way of reforming natural philosophy and wresting control of the scientific endeavour from the clutches of the Royal Society. Sir Joseph Banks (1743–1820), president of the latter for forty years, had once jeered to Sir John Barrow (1804–45) at the Admiralty, 'I see plainly that all these new-fangled associations will finally dismantle the Royal Society, and not leave the old lady a rag to cover her.'[37] In this sense the Royal Society, for reformers in general, was analogous to the prevailing national constitution and needed a radical overhaul. Indeed, Banks's prophecy had become fact by 1831 when the Astronomical Society, now the Royal Astronomical Society, became the leading technical body called upon to offer the government advice. Herschel implied as much when he chastised the recent attempt of Davies Gilbert (1767–1839) to ensure his presidency of the Royal Society would be succeeded by the Duke of Sussex: '[If] Mr. Gilbert has neither the capacity to conduct the business of the Society properly, nor the spirit to hold his station within it with dignity, he ought at least not to be suffered to imagine that he can hand it over like a rotten borough to any successor be his rank or station what they may by his *ipse dixit*' – that is, by Gilbert's own assertion without any authority or proof.[38] However, despite his attempt to pre-empt Gilbert's machinations by allowing his name to stand for nomination, Herschel lost the election to the Duke.

The composition and techniques adopted by the Boundary Commission for parliamentary reform were designed to distance politics from constitutional change via the use of scientific principles. In this sense, political reform was an inductive project, utilising precision instruments, with results reduced by mathematics; the method and results of the Commission were all visible and accountable. This was exactly the way the Astronomical Society advised science should be conducted and was central to Herschel's recent manual on the scientific method, the *Preliminary Discourse*, possibly timed to aid the process.[39]

The leader of the Boundary Commission Committee, the mathematician and army surveyor Thomas Drummond (1797–1840), was elected a fellow to the Astronomical Society in 1830. His election was probably promoted by the head of the Ordnance Survey, Thomas Frederick Colby (1784–1852), the head of the Royal Navy's Hydrographic Department, Francis Beaufort (1774–1857), and the Whig MP for Lichfield, John Wrottesley (second Baron Wrottesely, 1798–1867), all founding and leading members of the Society. The latter two also sat on the Boundary Commission along with another member of the Society, Richard Sheepshanks (1794–1855), son of a Leeds textile mill owner, astronomer, and Trinity graduate.

What makes the reform act so interesting is the way a new notion of science was harnessed to a political question to dilute the enormous opposition to the bill. The actual work of the Commission was carried out by an army of engineers and surveyors orchestrated by Drummond and his close associates from the Ordnance Survey such as Colonel Robert K. Dawson (1798–1861), who also sat on the Commission and was also based at a military institution (Woolwich and the Corps of Royal Engineers). The *Morning Post* feared the outcome of these trigonometrical practices and mathematical calculations in reshaping Britain's hallowed ancient constitution.[40]

For High Anglicans, this mathematically informed political reform was yet another example of Satan waving his wand and severing divine intelligence with an emphasis upon fallen human reason. Historian Robert Saunders recently summed up the true significance of the reform act: 'Implicitly and sometimes explicitly, 1832 had established the principle that government was a human construct, to be remodelled at the will of its creators. It was here, perhaps, that the reform act was most transformative. It is in this sense, not in the number of voters it created, that it marks an important moment in Britain's democratic history.'[41] Reform, in general, was opposed by critics as being an attack on age-old theological principles legislated by God's government – not human reason – and communicated via the established church. This was certainly the view of Whewell, for example, based at Trinity, and the basis of his attempt to define science within such boundaries; for Whewell, there was an organic relationship between man and divine knowledge – just as there was an organic relationship between the English Church, the state, and governance.[42]

William Van Mildert (1765–1836) – bishop of Durham, arch critic of the French Revolution, and one of the founders of the Anglican University of Durham (significantly, also founded in 1832) – told the House of Lords that year that he 'had attended to what was called the march of intellect, and . . . found abroad a restless disposition – a love of innovation – a wish to destroy institutions because they were ancient'.[43] The High Church quarterly publication, The British Critic, expressed the same sentiments that year: 'The principles . . . of the French Revolution, are again stalking abroad among us under other names; and it is the same Devil, who is still seeking what he can devour.'[44] The Church and constitution were quickly being dismantled – first by revolutions abroad and then by Catholic emancipation and the repeal of the Test and Corporations Acts, followed by the Great Reform Act.

Having been promoted by the scientific reforming element for, but defeated as, the next president of the Royal Society in 1830, Herschel was next utilised to verify one of Drummond's more sensitive surveying calculations.[45] The country was divided into nine pairs of commissioners, each of whom was assigned to assess twenty to thirty distinct districts, aided by several surveyors. The aim was, on paper, simple: to produce boundaries with at least three hundred households valued at ten pounds (calculated from tax returns or parish rates or, where necessary, actual valuation). In practice, this was a difficult task with many of the ancient district boundaries unclear and requiring the commissioners to expand (or shrink) districts – making these choices extremely politically sensitive.[46] It was this second aspect involved in the forging of new boundaries – factoring in what each household paid in tax – that called for Herschel and others to step in.[47]

When it came to the algorithm used to calculate the population of 120 of England's smallest boroughs based on their number of houses and the assessed taxes each paid, Drummond faced a very credible critic: Jonathan Frederick Pollock (1783–1870), the Tory MP for Huntingdon who condemned Drummond's calculation. Drummond added two figures together, whereas Pollock argued that they should be multiplied. Drummond's calculation thus worked like this: the number of houses in each borough divided by the average number of houses in each borough, plus the amount of assessed taxes paid in a borough divided by the average number of assessed taxes for all 120 boroughs. Pollock was a perfect adversary for Drummond: educated at St Paul's School and Trinity, he had graduated as Cambridge's first Smith's prizeman and senior wrangler in 1807 and was a leading Banksian and fellow of the Royal Society since 1816.

Banks's old order – a mirror of the national political stage – was not going down without a fight. If the opposition could not beat the reformers using divine arguments, they could take them on with their own mathematical techniques. Pollock and other mathematicians unleashed a torrent of letters to newspapers and pamphlets on the matter. In parliament, Pollock was supported against Drummond by the staunch conservative, anti-reformer, and the Secretary to the Admiralty, John Wilson Croker (1780–1857), who also opposed reform of the Hydrographic Office and disliked Drummond's ally Beaufort. Drummond was mortified by Pollock's assault and rushed to gain authority for his algorithm from his former mathematics mentor at Woolwich, Peter Barlow (1776–1862), as well as the Astronomer Royal, George Biddell Airy (1801–92), another first Smith's prizeman and senior wrangler from Trinity. He also approached Professor William Wallace (1768–1843) at Edinburgh University, formerly mathematics teacher at the Royal Military College at Marlow (later renamed Sandhurst). However, the real coup for Drummond was gaining the backing of Britain's leading man of science, Herschel. All four men Drummond turned to for support against Pollock's attacks were founding members of the Astronomical Society of London.[48]

During this same period, the Society took over from the Board of Longitude production of *The Nautical Almanac*. Here the aim was, as Herschel put it in 1825, to divest nautical computation 'of those technicalities which are not only puzzling to learn, but which really act as obstacles to its improvement by placing it in the light of a craft and a mystery, and obstructing that free access of scientific enquiry to all its processes'.[49] The 1833 *Nautical Almanac* was now the product of the forty, largely middle class and reform-minded members of the Astronomical Society, rather than the conservative bureaucratic establishment of the Board.

All this political turmoil took a toll on Herschel, and two years later he escaped the turbulence of the metropolis and landed at Table Bay in the British colony of Cape Town, South Africa. Margaret Herschel wrote to her brother's wife in 1835: 'I do think that Herschel would have been killed, had he remained in England during these troublesome times.' That trouble also included a nasty squabble between Herschel's friend and fellow member of the Astronomical Society, South, and the instrument maker Edward Troughton (1753–1835), who only a few years earlier had been important in getting South elected to the Royal Society of Edinburgh.[50]

However, it was Herschel's Unitarian and reforming tendencies that worried Margaret most. She told her brother, 'We must have our Herschel a decided & recognised Christian, "a city set on a Hill."' However, she went on, 'perhaps it must be attempted on our knees & in our conduct & conversation than by our arguments – *these* will please & delight afterwards. Probably (& I know it) he has read more of controversial works, & the writing of the Fathers than we have, but he confesses, that he was not accustomed to watch the life of a humble & sincere Christian living & rejoicing in their God'.[51] Despite her best efforts, Herschel struggled to conform to her wishes, and a nagging disapproval permeates her correspondence.

The Creation of Knowledge

There was no unified 'Cambridge network' presiding over the constitution of knowledge during this period. For Whewell, the scientific study of God's creation required a moral demeanour; his ideal natural philosopher was someone with a devout soul, a pious person like the genius Newton. His close Trinity friend Julius Hare (1795–1855) agreed and preached, 'Only through Faith, and by that patience and perseverance which a firm Faith alone can give, has knowledge ever been increast [sic] and exalted.'[52] Those who made great discoveries – such as Copernicus, Kepler, and especially Newton – possessed the religious disposition necessary to reflect, albeit dimly, the mind of the deity.[53] Even here, however, 'fundamental ideas' of space, time, force, and matter would still be 'far too obscure and limited to be regarded as identical with the Divine Idea'.[54] Whewell was much closer to former Trinity colleagues like Hare than Herschel.[55]

This perspective made the recent publication of Francis Baily (1774–1844), Herschel's like-minded friend, former stockbroker, actuary, Unitarian, and founding member of the Astronomical Society, especially damning. Baily wrote promoting the role of John Flamsteed (1646–1719) in the discovery of gravity, mischievously – but powerfully – underlining Newton's suspected Unitarianism and Edmund Halley's alleged atheism in his account.[56] How could such un-heavenly minds be so embedded in Anglican education and harnessed in support of moral standards? It is unclear how much Hare and Whewell knew about Newton's actual religious proclivities; for example, Newton privately thought 'the central Christian doctrine of the Trinity was a diabolical fraud'.[57] Certainly by the time the Scottish controversialist and fierce Presbyterian, David Brewster (1781–1868) waded in with his second history of Newton in 1855, it became clear that Newton had spent a great time researching alchemy and was a staunch anti-Trinitarian. However, the view of Newton as a genius in the Christian tradition continued to remain taken for granted – a view that continues to the present day.[58]

When Baily, one of Herschel's closest confidents, in his *An Account of the Rev. John Flamsteed*, suggested that if anyone fit Cambridge's Christian credentials it was the devout Anglican Flamsteed and not Newton, the suggestion was designed as a heavy blow to Newton's legacy and, indeed, traditional Anglican epistemology: 'All his [Flamsteed's] letters breathe a spirit of piety and resignation to the will of heaven.' It was not 'to be expected', Baily claimed, 'that the loose and irreligious conduct of Halley, both in his conversations and in his principles, could be at all congenial to a mind constituted like Flamsteed's'.[59] In this reading, only Flamsteed had a truly religious soul. This humorous but venomous sarcasm was designed to hit at the heart of the Anglican establishment. Whewell's vision of devout true science was exploded by Baily's narrative of how far great discoverers fell short of it, and indeed violated it. Baily had put the institutional alliance between Newtonianism, Christian epistemology, and the Church establishment on trial.

The detection of gravity was made possible only by the quantity and quality of Flamsteed's raw materials (observations), not Newton's ideas. Flamsteed's work dictated the value of the finished manufactured good, be it material or social truth. The system of producing knowledge from that data, like the Soho Foundry, was just the logical workings of the human mind that ground out the knowledge. Herschel told Whewell exactly what he thought of his philosophy in a letter written in August 1840: 'You are too apriori rather for me – as soon as one has worked ones way up to a general law', he scoffed, 'you come cranking in and tell me it is a fundamental idea innate in everybody's mind – a necessary truth or very likely to prove such in one thousand years hence'.[60] Many, like Herschel, found this hybrid Kantian aspect of Whewell's metaphysical philosophy abhorrent. For Whewell, fundamental ideas such as space, time, and number underpinned geometry and arithmetic, with force and matter lying at the heart of the mechanical sciences just as verticality was at the core of architecture. A few months later, while working on his review of Whewell's *Philosophy of the Inductive Sciences*, Herschel further warned, 'I fear talk as we might, or write as we may we shall never convert each other. We are like two staunch politicians Tory & Radical who agree in love of country and whom a thousand delightful associations keeps from tearing each other's eyes.'[61]

Herschel remained hostile to Whewell's metaphysical argument all his life. Whewell's nemesis, John Stuart Mill (1806–73), modelled his philosophy of induction on Herschel's *Preliminary Discourse*. It was human sensations excited by matter that made manifest external objects, and this was independent of any divine inspiration. It was human 'greater logic', to use Herschel's phrase, that interpreted them.[62] Herschel had no time for the German metaphysical philosophy that Whewell so loved and confessed to Margaret that he 'tried unsuccessfully to understand Immanuel Kant and J. G. Fichte' during their earlier voyage to the Cape.[63] In the same letter he told her that he ultimately found their philosophy nothing more than a snowball that had been rolled and rolled till it could not get any larger: 'Then came the sun &c &c &c and their universe turned into a puddle of dirty water.'[64]

Education and the Royal Mint

Whewell continued all his life to fight Cambridge University reform. A stern test of his resistance came when a Royal Commission was set up during the 1850s to inquire into the Cambridge and Oxford curriculum and examination system. Now the Master of Trinity, Whewell was furious, believing any reform should be done internally, not externally. In particular, he was resistant to any attempt to shrink the autonomy of the colleges. The idea, for example, of diluting the number of fellowships at Trinity and St John's to provide a university-wide fund was detestable to Whewell. This stance would bring him into conflict with the Trinity trio Adam Sedgwick (1785–1873), George Peacock, and Joseph Romilly (1791–1864), but also, significantly, with Herschel – all of whom sat on the reforming commission. Herschel was desperate to eradicate old, frequently medieval, university practices and replace them with contemporary liberal ones.

For Whewell, deluded utilitarian, calculating political reformers like Herschel seemed to be running rampant, seeking radical institutional change everywhere they went. He assured Hare he was doing all he could 'of the object you speak of – diminishing the influence of private tutors'.[65] In an earlier curriculum reform of 1848, Whewell had already fallen out with several of his college and university colleagues, especially the famous private tutor and 'senior-wrangler maker' William Hopkins (1793–1866), over the role of private tutors. This debate was aired again in the subsequent Royal Commission's investigations of curriculum teaching at Cambridge, leading to tension between the Trinity master and the reforming commissioners. Whewell firmly told them to back off, that private coaching 'enfeebles the mind and depraves the habits of study'.[66]

By May 1855, Whewell was bewildered by the actions of his friends and colleagues on the reforming commission and the general feeling for reform. He later told his wife, Everina (née Ellis), 'It is a very disgusting and disheartening subject. The bad faith of the ex-commissioners and the democratic frenzy which prevails in the University make it difficult to act or write with any hope of doing good; but I do not think I can help taking up my testimony against what they are doing.' Whewell concluded that he felt 'disgust, grief, and hopelessness' from 'treatment at the hands of friends' – and none more so than Herschel, who carried the most clout.[67] However, the university reform bill proposed in parliament was nothing like the Commissioners report, a change in the original university reform attempt that made both Peacock and Herschel furious. Peacock told the latter that Whewell was 'arbitrary and despotic'.[68]

Besides attempts at reforming Cambridge and Oxford, Herschel was also instrumental in introducing a state system of education to the Cape Colony, including a strong utilitarian curriculum and the assurance that decent salaries would be paid to the British teachers appointed (chosen by Herschel). He also lobbied government over their treatment of Andries Stockenström (1792–1864), the recently sacked Commissioner General for the eastern South African Province. Stockenström had become very unpopular with settlers keen to expand colonisation when he legislated policies preventing expansion into Xhosa lands. Herschel actively gave him political support and offered him financial assistance, which he declined.[69]

Along with attempting to reform education in Cambridge and the Cape, Herschel was also put in charge of reorganising the Mint as its penultimate Master. Two years after the Royal Commission report on the 'Constitution, Management, and Expense of Our Royal Mint' was published (1848), Herschel was put in charge of implementing its findings, which meant a complete overhaul of its medieval structure along rational lines; dissolving the contract system and replacing the work of minting by a full-time government officer. The five private moneyers, who had previously been responsible for minting coins, were given notice but generously compensated in 1851.[70] The traditional and customary procedures of the Mint were to be streamlined and made more efficient. Another major task for Herschel in this role was mapping the international supply and demand for gold and silver and predicting its availability for the next twenty years. The attempt to reform the Mint along rational grounds had taken its toll on a now aging Herschel. In 1857 he was replaced by the Professor of Chemistry at University College London, Thomas Graham.[71]

The year 1857 can be considered the moment Herschel decided to withdraw from public life: he resigned as Master of the Mint, from the Royal Astronomical Society, and – two years later – from the Royal

Society. Having been a promoter of the French metric system of metrology most of his life, he ironically spent his final years fighting against Britain and India adopting the system.[72] Throughout his life and career, Herschel's view of human intelligence and its rational functioning had a direct and sustained influence on his political stances, putting him at odds with even some of his closest and oldest colleagues, such as Whewell, and making him a proponent of reform in the government, education, universities, and churches.

Notes

1 The most public attempt to build a form of human intelligence into a machine was by Babbage, see S. Schaffer, 'Babbage's Intelligence: Calculating Engines and the Factory System', *Critical Inquiry*, 21.1 (1994): 203–27. For mechanistic objectivity in terms of visualised depictions, see L. Daston and P. Galison, *Objectivity* (Princeton: Princeton University Press, 2007), ch. 3.

2 J. Herschel, *A Preliminary Discourse on the Study of Natural Philosophy* (Chicago: Chicago University Press Reprint, [1830] 1987), 18–19.

3 Herschel, *Preliminary Discourse*, 19. Babbage had in his unpublished 'The Philosophy of Analysis' (1810s) demarcated symbolic algebra from its traditional application to simply arithmetic, see J. M. Dubbey, *The Mathematical Work of Charles Babbage* (Cambridge: Cambridge University Press, 1978), 100.

4 W. J. Ashworth, 'John Herschel, George Airy, and the Roaming Eye of the State', *History of Science*, 36.2 (1998): 151–78.

5 See Chapter 2.

6 H. J. Rose to W. Whewell, 5 October 1822, Trinity College (TC): Add.Ms. a.211/134.

7 J. Herschel, 'Calculating Machinery', *The Times*, 19 August 1828. About a month earlier Herschel had advised Joseph Clement to reduce 'the rapid rate of expenditure of the money on Charles Babbage's calculating machine'. Herschel to Joseph Clement, 24 July 1828, see M. J. Crowe, D. R. Dyck, and J. R. Kevin (eds.), *A Calendar of the Correspondence of Sir John Herschel* (Cambridge: Cambridge University Press, 1998), 95.

8 W. J. Ashworth, 'Memory, Efficiency, and Symbolic Analysis: Charles Babbage, John Herschel, and the Industrial Mind', *Isis*, 87.4 (1996): 629–53. For Babbage, see K. Lambert, *Symbols and Things: Material Mathematics in the Eighteenth and Nineteenth Centuries* (Pittsburgh: University of Pittsburgh Press, 2021), 73–76.

9 The development of symbolic algebraic logic was subsequently spearheaded by Augustus De Morgan and George Boole. For the former see, for example,

Formal Logic; or, the Calculus of Inference, Necessary and Probable (London, 1847); and for the latter, *The Mathematical Analysis of Logic: Being an Essay towards a Calculus of Deductive Reasoning* (London, 1847); and, especially, *An Investigation of the Laws of Thought, on which are founded the Mathematical Theories of Logic and Probabilities* (North Charleston: CreateSpace, [1854] 2016 reprint). Boole wrote: 'There is not only a close analogy between the operations of the mind in general reasoning and its operations in the particular science of Algebra, but there is to a considerable extent an exact agreement in the laws by which the two classes of operations are conducted,' 4. Boole had the patronage of Edward Bromhead, a member of the Cambridge Analytical Society, and the Cambridge mathematician, Duncan Farquharson Gregory. The latter had been educated in Scotland by the staunch analytical, William Wallace. See Lambert, *Symbols and Things*, 101–207.

10 D. Forbes, *The Liberal Anglican Idea of History* (Cambridge: Cambridge University Press, [1952] 2006), 36–37, 40–41, 58. For the ramifications see W. J. Ashworth, *The Trinity Circle: Anxiety, Intelligence, and Knowledge Creation in Nineteenth-Century England* (Pittsburgh: University of Pittsburgh Press, 2021), epilogue.

11 Ashworth, *The Trinity Circle*.

12 The most well-known critique of the mathematics taught at Cambridge came from the Scottish mathematician John Playfair, who made a scathing attack on both Oxford and Cambridge in an 1808 review of Laplace's *Mécanique céleste* for the *Edinburgh Review*; see Amy Ackerberg-Hastings, 'Analysis and Synthesis in John Playfair's "Elements of Geometry"', *British Journal for the History of Science*, 35.1 (2002): 43–72, on 67 and 70.

13 J. Herschel and C. Babbage, 'Preface to *Memoirs of the Analytical Society*', in M. Campbell-Kelly (ed.), *The Works of Charles Babbage* (London: W. Pickering, 1989), vol. 1, 37–60.

14 L. J. Snyder, *The Philosophical Breakfast Club: Four Remarkable Friends Who Transformed Science and Changed the World* (New York: Broadway Books, 2010). This confusion seems to stem from W. F. Cannon's earlier attempt to place these men within a distinct Cambridge network; see Cannon's, 'Scientists and Broad Churchmen: An Early Victorian Intellectual Network', *Journal of British Studies*, 4.1 (1964): 65–88.

15 Ashworth, 'Memory, Efficiency, and Symbolic Analysis', 629 and 632. J. L. Richards shows this clear demarcation between Whewell and Herschel in her *Mathematical Visions: The Pursuit of Geometry in Victorian England* (London: Academic Press, 1988), 24–25. Herschel's ambiguous religious views were probably informed by his father's Lockean deism, and is alluded to in G. Buttmann, *The Shadow of the Telescope: A Biography of John Herschel* (London: Lutterworth Press, 1974), 172–73.

16 Étienne Bonnot de Condillac, *Logic, or the First Development of the Art of Thinking* (Paris, 1780), reprinted in *Philosophical Writings of Étienne*

Bonnot de Condillac, trans. Franklin Philip with Harlan Lane (London: L. Erlbaum Associates, 1982), 414. The link between Condillac, Lagrange, Laplace, and Lacroix are best described in M. Panteki, 'Relationships between Algebra, Differential Equations, and Logic in England, 1800–1860' (PhD diss., University of Middlesex, 1991), 59–60; and Panteki, 'French "Logique" and British "Logic": On the Origins of Augustus De Morgan's Early Logical Inquiries, 1805–1835', in D. M. Gabbay and J. Woods (eds.), *Handbook of the History of Logic*, vol. 4, British Logic in the Nineteenth Century (Amsterdam: North Holland, 2008), 387–90, 396–97.

17 For the use of algebraic analysis outside of Cambridge, see, for example, A. D. D. Craik, 'Calculus and Analysis in Early 19th-Century Britain: The Work of William Wallace', *Historia Mathematica*, 26 (1999): 239–67.

18 A. Rogers to Herschel, 5 November 1808, 6 January 1809, 6 January 1812, and 1 January 1813, Herschel Papers, Royal Society.

19 T. Milnes, 'Coleridge's Logic', in Gabbay and Woods (eds.), *Handbook of the History of Logic*, 37–74, on 47 and 56.

20 William Hamilton is quoted in R. Olson, *Scottish Philosophy and British Physics: A Study in the Foundations of the Victorian Scientific Style* (Princeton: Princeton University Press, 1975), 21. For Herschel, see his scribblings in his copy of J. Robison, *Elements of Mechanical Philosophy, being the Substance of a Course of Lectures on that Science*, 2 vols. (Edinburgh, 1804), vol. 1, Sidney Ross Book Collection, Albany, NY.

21 Anonymous, 'Review of A Treatise on Plane and Spherical Trigonometry. By John Bonnycastle', *Eclectic Review* (January 1808): 52–59, on 58–59. This was a favourable review of Bonnycastle's work since it remained applied mathematics and was clearly demonstrated.

22 Ashworth, 'Memory, Efficiency, and Symbolic Analysis', 637; and *The Trinity Circle*.

23 J. Herschel, 'Great Physical Theories', n.d., Herschel Papers, Harry Ransom Center, University of Texas at Austin (hereafter Herschel Papers), quoted in Ashworth, 'Memory, Efficiency, and Symbolic Analysis', 638–39.

24 J. Herschel, 'Travel Diary, 1809–1810', 22 August 1809 and 17 July 1810, 'Scraps of Philosophy', n.d., Herschel Papers. For Babbage's similar perspective see Schaffer, 'Babbage's Intelligence'.

25 J. Herschel, 'Elements of Greatness', n.d., Herschel Papers.

26 Herschel, 'Scraps of Philosophy'.

27 J. Herschel, 'Mathematics', *Encyclopaedia Metropolitana* (1830); the latter is reprinted in S. S. Schweber (ed.), *Aspects of the Life and Thought of John Herschel* (New York: Arno Press, 1981), 434–59, on 434–35.

28 D. S. Evans et al. (eds.), *Herschel at the Cape: Diaries and Correspondence of Sir John Herschel, 1834–1838* (Houston: University of Texas Press, 1969), 334.

29 M. Herschel to Dr J. Stuart, 4 May 1837, in B. Warner (ed.), *Lady Herschel: Letters from the Cape, 1834–1838* (Cape Town: Friends of the South African

Library, 1991), 132–33. For Herschel's theological doubts and republican tendencies at Cambridge, see, for example, S. S. Schweber, 'John Herschel and Charles Darwin: A Study in Parallel Lives', *Journal of the History of Biology*, 22.1 (1989): 1–71, on 11–13.

30 Herschel to J. Whittaker, 2 July 1813, Whittaker Correspondence, St John's College; William Herschel to Herschel, 8 November 1813, quoted in C. A. Lubbock (ed.), *The Herschel Chronicle: The Life Story of William Herschel and His Sister Caroline Herschel* (Cambridge: Cambridge University Press, 1933), 349–50; Schweber, 'John Herschel and Charles Darwin', 11–13.

31 Evans et al., *Herschel at the Cape*, 32 and 240.

32 Evans et al., *Herschel at the Cape*, 37.

33 Herschel, 'Scraps of Philosophy'.

34 W. J. Ashworth, 'The Calculating Eye: Baily, Herschel, Babbage and the Business of Astronomy', *British Journal for the History of Science*, 27.4 (1994): 409–41.

35 For a concise overview of the unreformed political constituencies, see Eric J. Evans, *The Great Reform Act of 1832*, 2nd ed.(London: Routledge, [1983] 1994), 4–20.

36 M. Spycal, '"One of the Best Men of Business We Had Ever Met": Thomas Drummond, the Boundary Commission and the 1832 Reform Act', *Historical Research*, 90.249 (2017): 543–66.

37 Banks is quoted in J. A. Secord, *Visions of Science: Books and Readers at the Dawn of the Victorian Age* (Chicago: University of Chicago Press, 2014), 58.

38 Herschel is quoted in D. P. Miller, 'The Royal Society of London 1800–1835: A Study in the Cultural Politics of Scientific Organisation' (PhD diss., University of Pennsylvania, 1981), 352–53.

39 The scientific emergence of such an approach is best told in D. P. Miller, 'The Revival of the Physical Sciences in Britain, 1815–1840', *Osiris*, 2 (1986): 107–34. For the Astronomical Society, see Ashworth, 'The Calculating Eye' and Secord, *Visions of Science*, 87–89.

40 Spycal, 'One of the Best Men of Business We Had Ever Met', 551–52.

41 R. Saunders, 'God and the Great Reform Act: Preaching against Reform, 1831–32', *Journal of British Studies*, 53.2 (2014): 378–99.

42 Ashworth, *The Trinity Circle*.

43 William Van Mildert is quoted in Saunders, 'God and the Great Reform Act', 391.

44 *The British Critic* is quoted in Saunders, 'God and the Great Reform Act', 395.

45 For a succinct overview of the scientific structure during this period, see J. B. Morrell, 'Individualism and the Structure of British Science in 1830', *Historical Studies in the Physical Sciences*, 3 (1971): 183–204. For Herschel

and the 1830 election of the president of the Royal Society, see Miller, 'The Royal Society', 350–82.

46 Spycal, 'One of the Best Men of Business We Had Ever Met', 553–61; and J. F. M'Lennan, *Memoir of Thomas Drummond* (Edinburgh, 1867), 140–41.

47 For general background to the relationship debated between property value and taxpaying, see N. LoPatin-Lummis, 'The 1832 Reform Act Debate: Should the Suffrage be Based on Property or Taxpaying?', *Journal of British Studies*, 46.2 (2007): 320–45. The place of tax in political radicalism and reform during the early nineteenth century is neglected. See W. J. Ashworth, *Customs and Excise: Trade, Production, and Consumption in England 1640–1845* (Oxford: Oxford University Press, 2003), 319–40.

48 Spycal, 'One of the Best Men of Business We Had Ever Met', 562; M'Lennan, *Drummond*, 145–59. J. M. Rigg, revised by P. Polden, 'Pollock, Sir (Jonathan) Frederick, First Baron (1783–1870)', in *Oxford Dictionary of National Biography* (Oxford: Oxford University Press, 2004). For Wallace's analysis, see Craik, 'Calculus and Analysis in Early 19th-Century Britain', 256–57; and for Croker, see Miller, 'The Revival of the Physical Sciences', 114. For Herschel's response, see M'Lennan, *Drummond*, 148–60. It also seems Francis Beaufort from the Royal Navy's Hydrographic Department and fellow member of the Royal Astronomical Society also encouraged Herschel to aid Drummond. See Beaufort to Herschel, 10 January 1832 and Herschel to Beaufort, 13 January 1832, in Crowe, Dyck, and Kevin, *Calendar of the Correspondence of Herschel*, 129.

49 Herschel, 'On the Board of Longitude', n.d. (1825?) Herschel MS, St John's College. 'Report of the Committee of the Astronomical Society of London; Relative to the Improvement of the Nautical Almanac, adopted by the Council, 19 November 1830', reprinted in the *Nautical Almanac* (London, 1833), xii–xxii.

50 Warner, *Lady Herschel*, 84. It seems the heated South had also fallen out with another (optical) instrument maker, George Dolland. See Charles Babbage to John Herschel, 20 May 1821, in Crowe, Dyck, and Kevin, *Calendar of the Correspondence of Herschel*, 35. Herschel had been planning to escape to the Cape soon after his defeat in the contest to become president of the Royal Society. See Herschel to Fearon Fallows, 9 January 1831, in Crowe, Dyck, and Kevin, *Calendar of the Correspondence of Herschel*, 118.

51 Warner, *Lady Herschel*, 132–33.

52 J. C. Hare, 'Sermon IV: Power of Faith in Man's Natural Life', in J. C. Hare (ed.), *The Victory of Faith and Other Sermons* (Cambridge, [1840] 2012 reprint), 32.

53 R. Yeo, 'Genius, Method, and Morality: Images of Newton in Britain, 1760–1860', *Science in Context*, 2.2 (1988): 257–84, on 272.

54 William Whewell, *On the Philosophy of Discovery* (London, 1860), 361–36.

55 Ashworth, *The Trinity Circle.*

56 For the role and inspiration of Unitarianism in Baily's thinking, see, especially, J. L. Richards, *Generations of Reason: A Family's Search for Meaning in Post-Newtonian England* (New Haven: Yale University Press, 2021).

57 R. Iliffe, *Priest of Nature: The Religious Worlds of Isaac Newton* (Oxford: Oxford University Press, 2017), 4.

58 R. Higgit, *Recreating Newton: Newtonian Biography and the Making of Nineteenth-Century History of Science* (Pittsburgh: University of Pittsburgh Press, 2007).

59 F. Baily, *An Account of the Rev. John Flamsteed, the First Astronomer Royal; Compiled from His Own Manuscripts and Other Authentic Documents Never before Published; Part II: John Flamsteed's British Catalogue* (London, 1835), xxi–xii. Baily left his house in Tavistock Square, upon the death of his sister, to Herschel.

60 Herschel to Whewell, 6 August 1840, TC: Add.Ms.a.207/45.

61 Herschel to Whewell, 17 April 1841, TC: Add.Ms.a.207/46.

62 Wettersten, 'William Whewell', 725. D. B. Wilson, 'Herschel and Whewell's Version of Newtonianism', *Journal of the History of Ideas,* 35.1 (1974): 79–97, on 81–82. L. J. Snyder, *Reforming Philosophy: A Victorian Debate on Science and Society* (Chicago: University of Chicago Press, 2006), 81–82, 97, 132 and 143. John Stuart Mill to Herschel, 1 May 1843, in Crowe, Dyck, and Kevin, *Calendar of the Correspondence of Herschel,* 271; and various correspondence in 1846, 322–24.

63 Herschel to Margaret Herschel, 9[?] July 1838, and Herschel to Margaret, 22 July 1838, in Crowe, Dyck, and Kevin, *Calendar of the Correspondence of Herschel,* 188–89.

64 Herschel to Margaret Herschel, 9[?] July 1838, and quoted in S. Ruskin, *John Herschel's Cape Voyage: Private Science, Public Imagination and the Ambitions of Empire* (Aldershot: Ashgate, 2004), 15. See also Herschel to Whewell, 6 August 1840 and 16 April 1841, in Crowe, Dyck, and Kevin, *Calendar of the Correspondence of Herschel,* 232 and 244.

65 Whewell to Hare, 31 March 1843, TC:Add.Ms.a.215/68.

66 Whewell is quoted in A. Warwick, *Masters of Theory: Cambridge and the Rise of Mathematical Physics* (Chicago: University of Chicago Press, 2003), 103.

67 Whewell to E. E. Affleck, 12 May 1855. See J. M. Douglas, *The Life and Selections from the Correspondence of William Whewell* (London, 1881), 440.

68 George Peacock to Herschel, 15 April 1855, 19 April 1855, and 24 May 1855, in Crowe, Dyck and Kevin, *Calendar of the Correspondence of Herschel,* 498–99.

69 For Herschel and University reform, see Buttmann, *The Shadow of the Telescope,* 189–90; and for the reform of education in the Cape Colony,

113–15. See also George Phipps (Lord Normanby) to Herschel, 23 February 1839, Herschel to George Thomas Napier, 28 March 1839, Herschel to Stockenstrom, 21 January 1840, Herschel to Lord John Russell, 2 February 1840, and Herschel to George Thomas Napier, 15 August 1840, in Crowe, Dyck and Kevin, *Calendar of the Correspondence of Herschel*, 200, 203, 220–21, and 232. The background to this brutal era is vividly told by J. Laband, *The Land Wars: The Dispossession of the Khoisan and AmaXhosa in the Cape Colony* (Cape Town: Penguin Random House South Africa, 2020).

70 Herschel to M. Herschel, 20 November 1850, in Crowe, Dyck, and Kevin, *Calendar of the Correspondence of Herschel*, 408.

71 Lord J. Russell to Herschel, 23 December 1850, in Crowe, Dyck, and Kevin, *Calendar of the Correspondence of Herschel*, 411. Buttmann, *The Shadow of the Telescope*, 178–79. G. P. Dyer, *The Royal Mint: An Illustrated History* (London: Royal Mint, 1986), 30–31; J. Craig, 'The Royal Society and the Royal Mint', *Notes and Records of the Royal Society of London*, 19.2 (1964): 156–67, on 164.

72 See, for example, Herschel to J. Taylor, 4 June 1863 and 11 June 1863, Taylor to Herschel, 19 June 1863, Herschel to G. Airy, 8 July 1863, Herschel to T. Maclear, 3 December 1863, Herschel to Taylor, 17 December 1863, Herschel, letter to *The Times*, 21 June 1864, Herschel to Airy, 27 September 1864, Herschel to the editor of *The Saturday Review*, 1865, Herschel to George Airy, 20 April 1866, Herschel to W. J. Herschel, 1867?, Herschel to R. Strachey, October 1867 and 5 November 1867, J. A. Hardcastle to Herschel, 30 April 1868, Herschel to J. Herschel (son), 9 February 1870, John Banks to Herschel, 2 May 1871, in Crowe, Dyck, and Kevin, *Calendar of the Correspondence of Herschel*, 584–85, 590, 599, 603, 608, 627, 635, 643–44, 649, 671, and 685. See also Chapter 9.

11

John Herschel's Methodology in the
Scientific Community

The *Preliminary Discourse on the Study of Natural Philosophy*,
which Herschel published in Dionysius Lardner's *Cabinet Cyclopaedia*
series in 1830, can be a difficult book to interpret.[1] As commentators
have emphasized, its content and the circumstances of its publication
indicate it is perhaps better understood not (or at least, not merely) as a
technical treatise on scientific inference and methodology but rather in the
tradition of "conduct manuals," a popular genre that offered readers
insight into how they might elevate and refine their character.[2] The
Preliminary Discourse invokes not just prescriptions for scientific practice
but also the epistemic and personal virtues of a good scientist and even
(perhaps especially) the merits of careful observation and the study of
science for the layman.[3] Science, Herschel writes, is exceptional in "filling
us, as from an inward spring, with a sense of nobleness and power which
enables us to rise superior to" the circumstances of our lives.[4] The
Preliminary Discourse was printed and bound inexpensively, widely sold,
and frequently reprinted.[5] Given its history, it seems likely that philosophers of science have been too quick in reading this work primarily
through the lens of its contributions to the epistemology of science.

That said, it remains the case that the second and largest part of the
Discourse was dedicated to a detailed study of scientific methodology,
one of the first and most significant of such treatises to have appeared
in decades in English and one that was cited – at least by other
philosophers of the day, such as William Whewell (1794–1866) and
John Stuart Mill (1806–73) – as having reinvigorated the exploration of
what would come to be called "the philosophy of science."[6] Discussions

among these three and others initiated a tradition that would lead, among other destinations, to the early positivism of Karl Pearson (1857–1936) and, across the Atlantic, to reflections on the scientific method by Charles Sanders Peirce (1839–1914). Peirce praised Herschel, Whewell, and Mill for together offering "some of the finest accounts of the methods of thought in science."[7]

Given the high esteem in which Herschel's colleagues held his *scientific* work, it certainly seems reasonable to assume his treatise on methodology would have had an impact on his peers. As other chapters in this volume attest, Herschel's name was a byword for scientific authority in astronomy, physics, geology, and beyond. Susan Cannon (1925–81) did not exaggerate when she wrote that through the middle of the nineteenth century "one answer to the question of how to be scientific ... might be, 'Be as much like Herschel as possible.'"[8] But since the philosophy of science was a novel endeavor, and scientists always have a contested relationship with philosophy, we should expect this impact to be diffuse and difficult to pin down.

This injunction to be "like Herschel" went beyond philosophy as well, being more than merely instruction to follow Herschelian rules for inductive inference. As Richard Bellon has noted, in this period the genres of popular manual of conduct and textbook of scientific methodology overlapped more than might now seem apparent. In Victorian Britain, Bellon writes, "scientific discovery was a moral process, not an isolated event," and "scientists deployed a long list of words to imbue favored scientific research with moral authority."[9] Scientific methodology was certainly a matter of proposing and evaluating putative scientific explanations in the correct way, following sound canons of experimentation, and so on, but it was also a question of cultivating the right kinds of epistemic virtues as a practicing scientist. A list of such virtues that Bellon draws from a collection of Herschel's published articles includes "*ardent, arduous, careful, diligent, disinterested, humble, impartial, indefatigable, industrious, laborious, methodical, painstaking, patient, perseverant, scrupulous,* and *zealous*."[10] In understanding Herschel's influence, then, it is important to look not only at scientific practice but also at scientific character.

The task of investigating Herschel's influence on scientists of his day thus begins to take on a different form. Herschel, with many others, was fashioning a new discipline, one whose relationship to science was still a matter for debate and the extent of which was taken to go far beyond what we would today call the philosophy of science (even shading off into practical advice). What we see in Herschel's relationship to other scientists is not a story of direct influence or the *Preliminary Discourse* being used as a manual for scientific practice (though several examples will come close). More accurately, Herschel's work is a central contributor to a shared context of mid-nineteenth-century scientific methodology, one that was elaborated across scientific disciplines by a host of figures. This context was shaped by the *Discourse*, to be sure, as well as in other ways, including correspondence and networks of personal connection and influence. Book reviews were also particularly important in this period, and this was no different in the case of Herschel's work. What I will call "Herschelian" philosophy of science, then, is both an element and a product of this broader context, both created by the *Discourse* and altered by the interaction between those ideas and Herschel's colleagues.

To explore the way this philosophy was put into practice, this chapter will briefly chronicle Herschel's relationship with three important figures from three different branches of natural science of the day: Charles Lyell, Charles Darwin, and Michael Faraday. In each, we see a different way in which Herschel's work informed their intellectual perspective and vice versa. We will also see that the *Preliminary Discourse* was taken seriously by figures who went on to exert massive influence in a variety of different disciplines, both in how these natural philosophers conducted their science and the manner in which they believed it was important to behave as scientists. After 1830, natural science would forever be – to at least some degree – Herschelian.

Lyell and the *Principles*

One reason a straightforward exploration of Herschel's "influence" is too simplistic to capture the landscape surrounding discussions of methodology in this period is encapsulated by his relationship with the eminent geologist Charles Lyell (1797–1875). Herschel was five years Lyell's senior, and the *Discourse* included an example drawn directly from the first volume of Lyell's famous *Principles of Geology* (also published in 1830; the next two volumes would follow in 1832 and 1833). Herschel was preparing his book while in regular contact with Lyell, and, as we will see, Lyell and Herschel's thoughts on the role of *veræ causæ* are so close as to be nearly indistinguishable. Herschel and Lyell thus form a perfect example of the kind of contribution that Herschelian philosophy of science made: Herschel's claims about scientific methodology both reflected and shaped one of the most important scientific works of the mid-nineteenth century.

Lyell's geology arose from a rich context of controversy between two schools of geological thought, which Herschel's close friend William Whewell would baptize as the "catastrophists" and "uniformitarians." According to the catastrophists, the evidence of geology – especially of massive upheavals and subsidence, broken and disarrayed geological strata, and so forth – demonstrates that the major features of the geological record have been shaped by massive, catastrophic geological events (possibly, for some, including the Noachian deluge) entirely different in kind from those that we witness today. The uniformitarians, on the other hand – represented initially by the work of James Hutton (1726–97), commonly read in the abridged version of Hutton's thought presented by John Playfair (1748–1819) – argued that the causes we see working around us at present, like erosion, earthquakes, subsidence, and so on, would be enough to produce all the geological changes we observe, if they were only given enough time to operate.[11] As Lyell summarized the history of the dispute:

> We have seen that, during the progress of geology, there have been great fluctuations of opinion respecting the nature of the causes to which all former changes of the earth's surface are referrible [*sic*]. The first observers conceived that ... there have been causes in action distinct in kind or degree from those now forming part of the economy of nature .... [Others, more recently,] infer that there has never been any interruption to the same uniform order of physical events. The same assemblage of general causes, they conceive, may have been sufficient to produce, by their various combinations, the endless diversity of effects, of which the shell of the earth has preserved the memorials.[12]

Lyell placed great stock in what he called the "undeviating uniformity of secondary causes" as a feature that develops in any sufficiently advanced scientific theory, implicitly consigning catastrophes like the biblical flood to the same dustbin with "demons, ghosts, witches, and other immaterial and supernatural agents."[13]

Herschel lights on precisely this aspect of Lyell's theorizing in the *Preliminary Discourse* when introducing his own understanding of a *vera causa*, a feature often taken to be central to Herschel's methodology (and about which more in the next section, when we turn to Darwin). Herschel argued at length that successful scientific progress is about building a stock of proximate causes known to exist and to act in the world around us. If we confirm their action in the proper way (showing, for instance, that they could give rise not only to the phenomena for which we developed them in the first place but significantly different other observed phenomena as well), then they receive the stamp of scientific legitimacy. "To such causes," Herschel writes, "Newton has applied the term *veræ causæ*; that is, causes recognized as having a real existence in nature, and not being mere hypotheses or figments of the mind."[14]

As he turned to providing examples, after a toy case in which he rejects the possibility that "plastic virtue" of the soil could be responsible for the formation of fossils (compared with the *vera causa* of the death of a shelled animal and the deposition of that shell on the seabed), Herschel raises a more complex case: the fact that the surface of the earth has cooled over geologic time. We do not, he claims, have a *vera causa* to which we can appeal in constructing an explanation of this fact, for we lack the requisite experience of a planet cooling from a molten state or the circulation of heat from the center of the earth to its surface. But what we do have, thanks to Lyell, is a *vera causa*–compatible explanation for the change in the distribution of land and sea over time. Lyell had demonstrated this explanation's bona fides, Herschel claims, with "the degradation of the old continents, and the elevation of new, being a demonstrated fact; and the influence of such a change on the climates of particular regions, if not of the whole globe, being a perfectly fair conclusion, from what we know of continental, insular, and oceanic climates by actual observation." In contrast to catastrophism, this means that "we have, at least, a cause on which a philosopher may consent to reason."[15] We do not yet have the evidence we need to say that Lyell has given us the sole, correct explanation for continental change – that will take more evidence and evaluation,

Herschel argues – but we do know that this is the *kind* of thing that could be legitimately admitted into scientific theorizing, because its action can be confirmed by direct inspection.

The affinity, then, between the approaches of the two men should be obvious. Precisely the feature of Lyell's geology that he believed distinguished it from its predecessors – its reliance on highly confirmed, observed causes at work in the world around us – was taken by Herschel to be one of the defining characteristics of acceptable scientific theorizing. The two corresponded regularly during the years immediately prior to the appearance of their books, and informal opportunities for the sharing of ideas were of course manifest in the tightly knit community of Victorian British science. Lyell's work would often be consulted for its (we might say Herschelian) methodological tenets, and Herschel was respected for his geological field work, the results of which he shared with Lyell.[16]

A few years later, during his time at the Cape in 1836, Herschel wrote to Lyell that the latter's approach to geology was exemplary for the future development of science. "I hope your example will be followed in other sciences," he told Lyell, "of trying what *can* be done by existing causes, in place of giving way to the indolent weakness of a priori dogmatism – and as the basis of all further procedure enquiring what existing causes really are doing."[17] As we have seen, Herschel's praise was not based on idle speculation about the nature of geology and Lyell's contribution to it. He told Lyell that he had read all three volumes of Lyell's *Principles* (more than 1,400 pages) no fewer than three times and offered an array of suggestions, comments, and critiques in domains as disparate as the geophysical, the geographical, and the botanical. Herschel's interest was not merely a matter of making obeisance to a renowned fellow scientist.

Lyell replied with a long letter of thanks, professing that "I may truly say that when the Royal Society voted me a medal for my book, I was not more gratified nor more encouraged than by your full and interesting comments which have given me a feeling of strength and confidence in myself, which will assist me in my future studies."[18] The following year, after his return to Britain, he wrote to Whewell, now explicitly describing his theory in the same terms that Herschel used.[19] He argued there that his critics, who accused him of naively taking on an over-broad uniformity of nature as an assumption rather than arguing for it, were mistaken. Rather, "the reiteration of minor convulsions and changes, is, I contend, a *vera causa*, a force and mode of operation which we know to be true."[20] Of course, an invocation of the notion of *vera causa* was not necessarily Herschelian; the concept was coined by Newton and also famously defended by Thomas Reid (1710–96).[21] But again, we can see Herschel and Lyell both working to reinforce the importance of Herschelian philosophy of science to geological practice in the 1830s. Admiration between them was mutual and based in no small part on a shared commitment to the same tenets of high-caliber scientific method.

Darwin and the *Origin*

In the letter Herschel sent to Lyell, quoted above, Herschel offered a long-winded concurrence with yet another feature of Lyell's argument (defending it in fact even more strongly than Lyell himself had in the *Principles*): his naturalistic account of the creation of new species. Though the nature of the laws governing the production of species remained obscure, there must assuredly be such laws, Lyell had asserted, as the process of extinction is clearly at work in the world around us and yet the number of species on the globe seems to have remained roughly constant over geologic time. We should value, Herschel wrote, Lyell's "unveiling a dim glimpse of a region of speculation connected with it where it seems impossible to venture without experiencing some degree of that mysterious awe" described in Virgil's *Aeneid* or Walter Scott's *The Monastery*. "Of course I allude," he clarified, "to that mystery of mysteries the replacement of extinct species by others."[22]

Lyell must have shared the letter with a young naturalist whom he knew was working on similar questions: Charles Darwin (1809–82) (Figure 11.1). Cannon noted that the mere existence of Herschel's speculation on naturalistic causes for the creation of species must have been liberating. The young Darwin, she writes, "was able to be almost completely insensitive to theological considerations concerning the origin of species."[23] Indeed, on the very first page of the *Origin of Species*, Darwin wrote that the biogeography of South America he

Figure 11.1 The young Charles Darwin, in a watercolor portrait painted by George Richmond in the late 1830s (Public domain)

observed on the *HMS Beagle* "seemed to me to throw some light on the origin of species – that mystery of mysteries, as it has been called by one of our greatest philosophers."[24] As Cannon has noted, "when an early Victorian writer says, for example, that 'one of the most profound philosophers and elegant writers of modern times' has stated such-and-such, the chances are good that the reference is to John Herschel."[25]

Herschel has proven a fruitful source for interpreting Darwin's project as it was laid down in the *Origin* – though not a source without its share of difficulties. The first of these is that a host of other nineteenth-century figures have been equally illuminating, including Whewell, Mill, and Auguste Comte (1798–1857). Finding Darwin's position within the landscape of methodological insight in this period is thus challenging.[26] One aspect is certain enough, though: Darwin took from Herschel's *Discourse* exactly the kind of ascription of personal virtues arising from the practice of science that Bellon has highlighted. As Darwin described his educational development in his *Autobiography*,

> During my last year at Cambridge I read with care and profound interest Humboldt's *Personal Narrative*. This work and Sir J. Herschel's *Introduction to the Study of Natural Philosophy* [the *Discourse*] stirred up in me a burning zeal to add even the most humble contribution to the noble structure of Natural Science. No one or a dozen other books influenced me nearly so much as these two.[27]

But did Darwin in fact learn anything substantive from Herschel "more complicated than," as Cannon provocatively puts it, "that it would be wonderful to be a scientist"?[28]

The answer to this turns on the interpretation of the structure of the *Origin*'s central argument. We know that Darwin read the *Discourse* (for the second time) in late 1838, just as he was crystallizing the theory of natural selection and beginning to think of it as a piece of public, presentable science.[29] And if one regards the *Origin* through Herschelian lenses, a consistent reading emerges. Darwin begins his work with three chapters delineating variation in domesticated plants and animals as well as in the wild, then arguing for the presence of a struggle for existence that leads to the production of far more offspring than can ever possibly survive. We can interpret this as roughly akin to establishing natural selection as a *vera causa*. As we saw in Herschel's example drawn from Lyell, this is a very minimal criterion: we have to show that natural selection operates in ways similar to other causes the action of which we have demonstrated in other contexts – in this case, things like domestic breeding and the tendency of "the lowest savages" to protect and reproduce their best animals over generations, thus "unconsciously" improving the quality of their stock over time.[30] Of course, these are phenomena *similar* to natural selection and not natural selection itself. But Herschel had made space for exactly this move, and he had done so in exactly the way that Darwin would. "If the analogy of two phenomena be very close and striking," Herschel wrote, "while, at the same time, the cause of one is very obvious, it becomes scarcely possible to refuse to admit the action of an analogous cause in the other, though not so obvious in itself."[31] Darwin's demonstration that these phenomena of variation and "selection" (whether in conscious breeding or unconscious herd-tending) are analogous to natural selection thus directly follows Herschel's playbook for the introduction of a *vera causa*.

But – as we also saw above – showing that something is a *vera causa* only makes it "a cause on which a philosopher may consent to reason." We then have more work to do: what Herschel called establishing the "adequacy" of a cause to produce the effects demanded of it. "Whenever, therefore, we think we have been led by induction to the knowledge of the proximate cause of a phenomenon," he argued, "our next business is to examine deliberately and *seriatim* all the cases we have collected of its occurrence, in order to satisfy ourselves that they are explicable by our cause."[32] Chapters 4 through 9 of Darwin's *Origin* give this kind of argument, describing how natural selection can produce different species, genera, and higher groups with wildly different characters as well as various traits of organisms that readers were likely to see as refutations of natural selection, like highly or precisely adapted organs (such as eyes), instincts (and other mental or cognitive capacities), sterile hybrids, and so forth.

And lastly – anticipating the development of consilience, for which Whewell would become well-known a decade later – Herschel argued that we must not rest content with establishing adequacy, since adequacy involves primarily testing against phenomena that we had in mind while developing our theory. We must then turn to "*extending its application to cases not originally contemplated* ... studiously varying the circumstances under which our causes act, with a view to ascertain whether their effect is general [and] pushing the application of our laws to extreme cases."[33] This is precisely what Darwin does in the last four chapters of the *Origin* (before the summary conclusion), where he shows that adopting an evolutionary perspective can shed light on geology, biogeography, taxonomy, morphology, embryology, and the existence of rudimentary organs.

A motivated reader, then, can analyze the structure of Darwin's *Origin* and see a theory designed precisely to satisfy Herschel's methodological precepts. Herschel's *Discourse* laid out the steps that one ought to take in the course of developing, proposing, and evaluating a new cause to be added to the stock of those available in natural science. Darwin offered arguments corresponding to each of these steps in the order and arrangement Herschel would have wanted. On this reading, Darwin's template or standard for what a piece of quality, publishable, public scientific theorizing should look like was drawn directly from Herschel's methodological maxims in the *Discourse*.

That said, there is contention concerning this view. At a fine-grained level, nearly any way of making it precise can be contested, as there are a number of incompatible ways to see how the various parts of Darwin's argument constitute a Herschelian story – which chapters contribute to which facets of the defense of natural selection.[34] But also, we might ask ourselves to what extent we have overestimated Darwin's own philosophical sophistication. Herschel's approach to the proposition and validation of *veræ causæ* is extremely subtle and has been subject to a variety of disagreements and misreadings in the philosophical community over the two centuries since Herschel set it down.[35] It is thus perhaps doubtful that Darwin indeed took from the *Discourse* the detailed structure for the introduction of a causal theory that commentators have argued is evident in the *Origin*. The uses that Darwin made of the *vera causa* concept across his various letters and notebooks, to take just one example, do not make it entirely clear what he took to be a *vera causa* or how he considered the many causes involved with natural selection to interact.[36] Speaking more generally, one of course need not have a sophisticated and consistent causal interpretation of natural selection to support evolutionary theory.[37] It is precisely these details that are difficult to discern in any particular case of "influence" and that make the indisputable proof of any such influence so hard to come by.

But for our purposes, an effort to use Herschel's philosophy of science – even if it were a heavy-handed and perhaps clumsy attempt, lacking the sophistication of a contemporary reading of the *Discourse* – still has Darwin drawing on the mid-nineteenth-century methodological context that was so strongly shaped by Herschel's work. The available circumstantial evidence, such as Darwin's having reread the *Discourse* just as he was attempting to structure his nascent thoughts about natural selection, offers us good reason to think we have an example of Herschel's direct influence on nineteenth-century scientific practice.

Another major problem with a rosy interpretation of the Herschel-Darwin relationship that merits mention here is Herschel's reaction to

Darwin's *Origin*. Herschel famously rejected the theory of evolution, though it can be hard to understand why, as he discussed it only very rarely in print. Darwin lamented in a letter to Lyell that "I have heard by round about channel that Herschel says my Book 'is the law of higgledy-pigglety.' – What this exactly means I do not know, but it is evidently very contemptuous. – If true this is great blow & discouragement."[38] Herschel's objection, however, need not have been made on methodological grounds. Herschel did not think any theory of evolutionary change could be considered adequate to produce evolutionary phenomena unless it could encompass a theory of the generation of the variation that is the raw material on which natural selection works. Since Darwin could not provide such a theory, his adequacy case for natural selection failed for want of evidence.[39] But this does not imply that Herschel believed Darwin had somehow misapplied his canons of good methodology; we do not have any documentary evidence that, for instance, Herschel thought the *Origin* was somehow non-scientific or badly argued. Insufficiently empirically supported science, Herschel might have said, is still science.

Darwin's critics, Herschel included, thus did not often frame their arguments against him in Herschelian-methodological terms. But his supporters did, at least occasionally, explicitly defend his work in this way. Writing in *The Geologist*, Frederick Hutton (1836–1905), who would go on to offer evolutionary accounts of the flora and fauna of New Zealand, argued it was self-evident that "natural selection is a 'vera causa,'" and, closing his article by citing the *Preliminary Discourse*, wrote that though "I know that it rests at present on presumptive evidence alone ... in the words of Sir John Herschel, 'are we to be deterred from framing hypotheses and constructing theories, because we meet with such dilemmas, and find ourselves frequently beyond our depths? Undoubtedly not.'"[40]

To sum up, there is a consistent reading of the *Origin* (and further evidence from the notebooks, Darwin's correspondence, and his peers) in which Darwin had in mind, in his presentation of and argument for natural selection, the structure contained in Herschel's *Discourse* for proposing, evaluating, and verifying a causal claim in natural science. This evidence is, however, somewhat mixed.[41] But even with this equivocal evaluation of the case, it seems clear Darwin took

Figure 11.2 Michael Faraday, painted in 1842 by Thomas Phillips (Public domain)

Herschel's philosophy of science very seriously, and Herschel's own appraisal of natural selection did not differ with it on methodological grounds.

Michael Faraday and Experiments on Light

One aspect of the relationship between Herschel and Darwin that makes it difficult to analyze is that Darwin did not spend much time directly discussing his philosophical debts. On the contrary, this is not a problem when we turn to the work of the renowned physicist Michael Faraday (1791–1867) (Figure 11.2). In 1832, Herschel had written to Faraday praising him for his recent experimental work.[42] Faraday wrote back that he was particularly touched:

> I have the more pleasure in receiving your commendation than that of another person – not merely because there are few whose approbation

I should compare with yours but for another circumstance. When your work on the study of Nat. Phil. [the *Discourse*] came out, I read it as all others did with delight. I took it as a school book for philosophers and I feel that it has made me a better reasoner & even experimenter and has altogether heightened my character and made me if I may be permitted to say so a better philosopher.

In my last investigations I continually endeavored to think of that book and to reason & investigate according to the principles there laid down.[43]

Once again, this was not idle praise for an important and influential colleague. A few months later, on March 25, 1833, Faraday found himself the de facto representative of the Royal Institution at a dinner in honor of the centenary of the birth of Joseph Priestley (1733–1804). Demurring that "I have no reason why I should be distinguished with this mark of your favor … except that of the absence of my superior," he went on to use the opportunity to reiterate, this time publicly, his praise for Herschel's *Discourse*:

For my own part I must acknowledge that I cannot but attribute much of my late experimental success to an endeavour to follow the candid method of investigation pursued by Priestley, and to apply the principles of philosophical logic which I found in Sir John Herschel's "Preliminary Discourse."[44]

Faraday took Herschel's work, then, to champion some of the same kinds of character traits – like "freedom of mind," "independence of dogma and of preconceived notions," and "observation of facts which result from natural causes working before us" – that he believed made Priestley's thought so valuable. The admiration was mutual; as Sydney Ross details, Herschel stood up for Faraday as an equal member of the scientific community and supported his (fiercely contested) membership in the Royal Society.[45]

Following the reconstruction by David Gooding of another episode in their relationship, we can explore the connection between Herschel and Faraday in greater detail.[46] In late 1845, Faraday announced he had discovered what has since come to be referred to as the Faraday effect: that the polarization plane of a beam of polarized light can be rotated under the influence of a magnetic field, proportional to the strength of the magnetic force. This is nearly a direct demonstration that light is, in fact, an electromagnetic phenomenon – a claim Faraday had long supported but not yet confirmed. Herschel wrote Faraday another congratulatory letter a few months later, after the public announcement of the discovery. This letter was tinged with a bit of scientific regret: Herschel had himself attempted to find evidence for the same phenomenon. "It is now a great many years ago," he wrote, "that I tried to bring this to the test of experiment," when he had attempted to use "a great magnetic display by Mr Pepys at the London Institution" to show the same kind of effect of magnetism on polarization. The experiment had failed. Herschel had no intention of questioning Faraday's priority – "for," he wrote, "though I may regret that I did not prosecute a train of enquiry which seemed so promising up to a decisive fact I consider it honour enough to have entertained a conception which your researches have converted into a reality" – precisely because of the crucial role that he gives to experiment in his newly developed philosophy of science.[47]

Gooding argues that this shows an interesting divergence between Herschel's philosophy and his own actual scientific practice. Whereas Faraday, Gooding writes, "never underestimated the difficulty of extracting the 'natural fact' from the phenomenal artefacts produced by his instruments," Herschel's approach to the question of magnetism and polarization

> reveals a discrepancy between his experimental practice and his methodology. According to the latter, experiment was primary. Thus, discoveries are awarded to the experimentalists who demonstrate them. Yet experiment was not actually as important to Herschel [in his scientific practice] as [his position expressed in the *Preliminary Discourse*] implies.[48]

If Herschel had successfully carried through his own precepts as laid down in the *Discourse*, he would have worked harder at repeating the hastily conducted experiment that he had performed using Pepys's battery (varying at least the two major possible explanations for failure, the battery's low charge and the medium in which the light was transmitted).

To see how Gooding's explanation might be supported, let's look at the way Herschel talks about the very idea of experiment in the *Discourse*. Collective and accumulated experience, he writes, is "the great, and indeed, only ultimate source of our knowledge of nature and its laws."[49] But experience can be generated in two different ways: observation, which simply consists of "noticing facts as they occur," and experiment, which results from "putting in action causes and agents over which we have control, and purposely varying their combinations, and noticing what effects take place."[50] Herschel wrote that he preferred to call these types of observations *passive observation* and *active observation*, to underline the idea that both, though they refer to different approaches and different states of mind, result in the end in the collection of facts from the world around us. But the inductive credentials of experiment across the history of science are impressive and are what distinguish it from passive observation. He draws out the case in a long analogy with testimonial evidence. We can either listen to the story that a witness tells us (often regretting later that we failed to pay attention to some important detail), or, by contrast,

> we cross-examine our witness, and by comparing one part of his evidence with the other, while he is yet before us, and reasoning upon it in his presence, are enabled to put pointed and searching questions, the answer to which may at once enable us to make up our minds.[51]

This grounds a substantial difference in power between experimental and observational sciences:

> Accordingly it has been found invariably, that in those departments of physics where the phenomena are beyond our control, or into which experimental enquiry, from other causes, has not been carried, the progress of knowledge has been slow, uncertain, and irregular; while in such as admit of experiment, and in which mankind have agreed to its adoption, it has been rapid, sure, and steady.[52]

These are strong words, especially coming from a scientist who has made his name in the family business of astronomy – exactly, one might think, the kind of department of physics where the phenomena are beyond our control. But it is the incorporation of astronomy as a branch of mechanics, the ability to refine our observational techniques in order to bring astronomy closer to the category of experimental science, and the ability to test its claims (especially in contemporary observational astronomy, see Chapters 3 and 6), that has enabled its recent and impressive advancement, Herschel claims.

Why does Herschel believe experiment has this privileged role in scientific practice? As he argued later, perhaps the most important reason for its superiority is the fact that "in nature, it is comparatively rare to find instances pointedly differing in one circumstance agreeing in every other; but when we call experiment to our aid, it is easy to produce them."[53] Experimentation thus gives us the ability to systematically vary the conditions that lead to a given phenomenon in the effort to confirm that a proposed cause is indeed the one responsible for it. And, as Gooding reconstructs the methodology found in Faraday's notebooks, this is exactly the way in which Faraday conceived of the nature and role of his experimental work. In investigating some phenomenon,

> it is impossible to predict the whole set of necessary conditions. These have to be learned by systematically varying the parameters in order to discover the relevant parameters .... Most of the work recorded in Faraday's laboratory *Diary* and (to a lesser extent) in his published *Researches*, is about this sort of problem-solving.[54]

In that sense, then, Faraday has out-Herscheled Herschel: Herschel didn't have the tenacity (or, one might demur, the time and access to high-quality equipment) to experiment further following his own guidelines for testing the effect of magnetism on light. But he did immediately recognize that the existence of that very tenacity – the fact that Faraday had adhered so precisely to the experimental method laid down in the *Discourse* – offered a clear confirmation of Faraday's legitimate priority (and virtue) in the discovery of the effect.

Faraday thus serves as perhaps the most direct example of Herschel's role in mid-nineteenth-century developments in scientific methodology. As between Herschel and Lyell, there was a deep and abiding mutual admiration between Herschel and Faraday, focused in no small part on precisely these questions of methodology, and, for Faraday as for Darwin, there was an explicit reliance on Herschel's *Discourse*. Faraday's admiration is even more clearly expressed though, and the reliance on the *Discourse* can be traced not only through oblique references and circumstantial evidence but also through Faraday's experimental practice itself and his discussion with Herschel on the physical effect that now bears his name.

Conclusion

As Susan Cannon has argued, John Herschel set the bar for what it meant to do science in the mid-nineteenth century. In one sense, this was due to the sterling example of his own scientific work. Herschel's astronomy served as the model, at the very least, for reasoning within the physical sciences and likely, at least implicitly, for sciences far beyond physics.

> In the England of the 1830s, "to be scientific" meant "to be like physical astronomy." To be quite specific, it meant "to be like John Herschel's extension of physical astronomy to the sidereal regions by his observations and then calculations of double-star orbits."[55]

But this exemplary role was also a result of his methodological and philosophical claims. As we have seen, his presentation of the precepts for introducing and proposing *veræ causæ* were influential on Lyell and Darwin; privately, his willingness to entertain a naturalistic explanation for the origin of species was important for Darwin as well; and his approach to experiment and observation, especially surrounding the persistent, systematic variation of the conditions under which a putative cause takes place, was a guiding principle for the experimental work of Faraday.

To close, I want to expand our view of the intellectual context to which Herschel contributed by returning to a point noted in the introduction. In addition to his methodological norms, Herschel advocated for a collection of epistemic virtues that could define what it meant to be a good *scientist*, not just to engage in good scientific practice. Of course, detecting the presence of these virtues in the works (or, perhaps better, in the lives) of nineteenth-century researchers is a challenge of a different order. But we can get some glimpses of what these qualities might look like for each of the three figures surveyed here.

To see the most explicit epistemic-virtue defense of the work of Lyell, we must briefly leave Herschel's writings and turn to Whewell's review of the first volume of Lyell's *Principles*, though the description there is entirely consonant with what we would find in Herschel's work. Because, Whewell writes, "a mass of knowledge has now been collected, most remarkable both in its quantity and its kind," we are finally capable of, "with a sagacity, perseverance, and success," profiting from "a fresh outbreak of the spirit of theorizing among our geologists."[56] Whewell wrote that "the book has in truth a higher character; for it is so constructed, that the reader may avail himself of Mr. Lyell's aid, his rich and pregnant observation, his sound and well-pondered comparison."[57] In short, Lyell's empirical grounding (in the body of carefully collected geological evidence) and his epistemic virtue ensure that even a speculative geological work will be worth our effort.

Turning to Darwin, Bellon notes that part of his triumph in convincing others of his new theory of evolution by natural selection was his having demonstrated precisely that he possessed such virtues. In addition to what might have been perceived as the rash theorizing present in a work like the *Origin*, he had already published, and would go on to publish, a host of other more methodical works on barnacles, orchids, earthworms, plant fertilization, insectivorous plants, and so forth.[58] Multiple commentators, including the botanist George Bentham (1800–84) and the chemist Charles Daubeny (1795–1867), stated publicly that this demonstration of virtue did much for their opinion not only of Darwin but of his theorizing more generally.

When Faraday linked Herschel's work to the types of desirable features he had seen in the paragon Priestley, he did so largely in epistemic-virtue terms: Priestley was unimpeded by preconceived notions and dogmas, which gave him the right kind of "freedom of mind" for scientific work. Faraday presumably had these sorts of criteria in mind when he wrote that having attempted to follow both Priestley's example *and* Herschel's *Discourse* was crucial to the quality of the experimental results he had been able to produce.

Both Herschel's standards for scientific methodology and his closely related model for scientific character and epistemic virtue were instantiated by some of the leading figures of the nineteenth-century scientific community in disciplines as diverse as geology, natural history, and (non-astronomical) physical science. These same figures, in turn, adopted, adapted, and advocated their own views, shaping the fertile environment of theorizing about science during this period. Whether the *Discourse* is read more narrowly as a work describing the epistemology of science and inductive inference, more moderately as a book about the kinds of epistemic virtues that practicing scientists needed to exemplify, or more broadly as a manual for good conduct both within and beyond the scientific community, it is clear that the history of science was indelibly marked by the change in philosophical perspective that took place during this period – a change of which Herschel was one of the primary architects.

Notes

1 John F. W. Herschel, *A Preliminary Discourse on the Study of Natural Philosophy*, 1st ed. (London: Longman, Rees, Orme, Brown, & Green, 1830). See Chapter 1 for a discussion.

2 James A. Secord, "The Conduct of Everyday Life: John Herschel's *Preliminary Discourse on the Study of Natural Philosophy*," in *Visions of Science: Books and Readers at the Dawn of the Victorian Age* (Chicago: University of Chicago Press, 2014), 80–106.

3 In order to avoid the complex landscape and shifting definitions of "science" and "natural philosophy" (as well as "scientist" and "natural philosopher") in this period, I have chosen to adopt "science" and "scientist" in roughly their contemporary sense here. The further intricacy that such a change in terminology would cause outweighs the increase in terminological accuracy. Herschel himself tended to use science, natural science, and natural philosophy as synonyms.

4 Herschel, *Preliminary Discourse*, 16, §12.

5 Secord, "The Conduct of Everyday Life," 81, 87.

6 William Whewell, "[Review of] A Preliminary Discourse on the Study of Natural Philosophy. By J. F. W. Herschel, Esq., M.A. of St. John's College, Cambridge," *The Quarterly Review*, 45.90 (1831): 374–407; John Stuart Mill, "Herschel's *Preliminary Discourse*," *The Examiner*, March 20, 1831; John F. W. Herschel, "Address of the President," in *Report of the Fifteenth Meeting of the British Association for the Advancement of Science* (London: John Murray, 1846), xl.

7 Karl Pearson, *The Grammar of Science*, 1st ed. (London: Walter Scott, 1892); Charles Sanders Peirce, *Collected Papers of Charles Sanders Peirce*, vol. 1: *Principles of Philosophy*, ed. Charles Hartshorne and Paul Weiss (Cambridge, MA: Harvard University Press, 1931), CP 1.29.

8 W. F. Cannon, "John Herschel and the Idea of Science," *Journal of the History of Ideas*, 22.2 (1961): 215–39, on 219.

9 Richard Bellon, "Sacrifice in Service to Truth: The Epistemic Virtues of Victorian British Science," in Emanuele Ratti and Thomas A. Stapleford (eds.), *Science, Technology, and Virtues* (New York: Oxford University Press, 2021), 17–36, on 18.

10 Bellon, "Sacrifice in Service to Truth," 18.

11 John Playfair, *Illustrations of the Huttonian Theory of the Earth* (Edinburgh: Cadell and Davies, 1802); Martin J. S. Rudwick, "Lyell and the Principles of Geology," *Geological Society, London, Special Publications*, 143.1 (January 1998): 1–15.

12 Charles Lyell, *Principles of Geology*, vol. I (London: John Murray, 1830), 75.

13 Lyell, *Principles of Geology*, 76.

14 Herschel, *Preliminary Discourse*, 144, §138.

15 Herschel, *Preliminary Discourse*, 147, §149.

16 Gregory A. Good, "John Herschel's Geology: The Cape of Good Hope in the 1830s," in Jed Buchwald and Larry Stewart (eds.), *The Romance of Science: Essays in Honour of Trevor H. Levere* (Cham: Springer, 2017), 135–50.

17 W. F. Cannon, "The Impact of Uniformitarianism: Two Letters from John Herschel to Charles Lyell, 1836–1837," *Proceedings of the American Philosophical Society*, 105.3 (1961): 307–8.

18 Cannon, "The Impact of Uniformitarianism," 311.

19 See also Whewell's review of Lyell's *Principles*, which demonstrates Whewell's familiarity with and approval of Lyell's work more generally (and about which more in the concluding section); William Whewell, "[Review of] *Principles of Geology; Being an Attempt to Explain the Former Changes of the Earth's Surface by Reference to Causes Now in Operation. By Charles Lyell, Esq. F. R. S. For. Sec. to the Geol. Soc., &c. In 2 Vols. Vol. I*," *The British Critic*, 9 (1831).

20 Katharine Murray Lyell, *Life, Letters and Journals of Sir Charles Lyell, Bart.* (London: John Murray, 1881), 2:3.

21 Thomas Reid, *Essays on the Intellectual Powers of Man*, ed. A. D. Woozley (London: Macmillan, 1941), 34–35. See also Rachel Laudan, "The Role of Methodology in Lyell's Geology," *Studies in History and Philosophy of Science*, 13.3 (1982): 215–49.

22 Cannon, "The Impact of Uniformitarianism," 305.

23 Cannon, "The Impact of Uniformitarianism," 302.

24 Charles Darwin, *On the Origin of Species*, 1st ed. (London: John Murray, 1859), 1.

25 She also gives the example of Mary Somerville. Cannon, "John Herschel and the Idea of Science," 218.

26 In this section, I follow portions of my previous analysis, though my opinion has shifted somewhat in the intervening years. Charles H. Pence, "Sir John F. W. Herschel and Charles Darwin: Nineteenth-Century Science and Its Methodology," *HOPOS*, 8.1 (2018): 108–40.

27 Charles Darwin, *The Autobiography of Charles Darwin, 1809–1882, with Original Omissions Restored*, ed. Nora Barlow (London: Collins, 1958), 67–68.

28 W. F. Cannon, "Charles Lyell, Radical Actualism, and Theory," *British Journal for the History of Science*, 9.2 (1976): 104–20, on 118.

29 Charles Darwin, 'Books to Be Read' and 'Books Read' Notebook (1838–51). CUL-DAR119, ed. Kees Rookmaker (http://darwin-online.org.uk/), fol. 4v.

30 Darwin, *On the Origin of Species*, 34.

31 Herschel, *Preliminary Discourse*, 149, §142.

32 Herschel, *Preliminary Discourse*, 165, §172.

33 Herschel, *Preliminary Discourse*, 167, §176.

34 Compare, for instance, my account in Pence, "Sir John F. W. Herschel and Charles Darwin" with that of M. J. S. Hodge, "Darwin's Argument in the Origin," *Philosophy of Science*, 59.3 (1992): 461–64.

35 Richard Yeo, "Reviewing Herschel's Discourse," *Studies in History and Philosophy of Science*, 20.4 (1989): 541–52; Marvin Paul Bolt, "John Herschel's Natural Philosophy: On the Knowing of Nature and the Nature of Knowing in Early-Nineteenth-Century Britain" (PhD Thesis, University of Notre Dame, 1998).

36 Ben Bradley, "Natural Selection according to Darwin: Cause or Effect?," *History and Philosophy of the Life Sciences*, 44.2 (2022): 13.

37 A fact made all the more salient by the fact that no such interpretation currently receives philosophical consensus. See Charles H. Pence, *The Causal Structure of Natural Selection* (Cambridge: Cambridge University Press, 2021).

38 Charles Darwin to Charles Lyell, "Letter 2575 – Darwin, C. R. to Lyell, Charles, [10 Dec. 1859]," December 10, 1859, www.darwinproject.ac.uk.

39 Pence, "Sir John F. W. Herschel and Charles Darwin," 130–35.

40 Frederick Wollaston Hutton, "Some Remarks on Mr. Darwin's Theory," *The Geologist*, 4 (1861): 188. Hutton's quote was from Herschel, *Preliminary Discourse*, 196, §208.

41 However much I may have argued in support of the contrary point in the past.

42 Michael Faraday, *Experimental Researches in Electricity* (London: J. M. Dent and Sons, 1914), secs. 1–2.

43 Michael Faraday to John F. W. Herschel, "Letter Faraday0623, from Michael Faraday to John Frederick William Herschel," November 10, 1832, https://epsilon.ac.uk/view/faraday/letters/Faraday0623. Accessed February 6, 2023.

44 William Babington et al., "Commemoration of the Centenary of the Birth of Dr. Priestley," *Philosophical Magazine*, 2.11 (May 1833): 390–91.

45 Sydney Ross, "John Herschel on Faraday and on Science," *Notes and Records of the Royal Society of London*, 33.1 (1978): 77–82.

46 David Gooding, "'He Who Proves, Discovers': John Herschel, William Pepys and the Faraday Effect," *Notes and Records of the Royal Society of London*, 39.2 (April 30, 1985): 229–44.

47 John F. W. Herschel to Michael Faraday, "Letter Faraday1783, from John Frederick William Herschel to Michael Faraday," November 9, 1845, https://epsilon.ac.uk/view/faraday/letters/Faraday1783. Accessed February 6, 2023.

48 Gooding, 'He Who Proves, Discovers,' 231.

49 Herschel, *Preliminary Discourse*, 76, §67.

50 Herschel, *Preliminary Discourse*, 76, §67.

51 Herschel, *Preliminary Discourse*, 77, §67.

52 Herschel, *Preliminary Discourse*, 77, §67.

53 Herschel, *Preliminary Discourse*, 155, §156.

54 Gooding, 'He Who Proves, Discovers,' 234.

55 Cannon, "John Herschel and the Idea of Science," 238.

56 Whewell, "[Review of] Principles of Geology," 180, 184.

57 Whewell, "[Review of] Principles of Geology," 186.

58 Bellon, "Sacrifice in Service to Truth," 30–31, Richard Bellon, "Charles Darwin Solves the 'Riddle of the Flower'; or, Why Don't Historians of Biology Know about the Birds and the Bees?," *History of Science*, 47.4 (December 2009): 373–406.

Further Reading

Chapter 1: John Herschel: A Biographical Sketch

Cannon, S., "John Herschel and the Idea of Science," *Journal of the History of Ideas*, 22 (1961), 215–39.

Crowe, M. J., D. R. Dyck, and J. R. Kevin (eds.), *A Calendar of the Correspondence of Sir John Herschel*, Cambridge, Cambridge University Press, 1998.

Ferguson, W. T., and R. F. M. Immelman, *Sir John Herschel and Education at the Cape, 1834–1840*, Cape Town, Oxford University Press, 1961.

Lubbock, C., *The Herschel Chronicle: The Life-Story of William Herschel and His Sister Caroline Hershel*, Cambridge, Cambridge University Press, 1933.

Chapter 2: The Mathematical Journey of John Herschel

Ashworth, W. J., "Analytical Society," in *Oxford Dictionary of National Biography*, Oxford, Oxford University Press, 2021. www.oxforddnb.com/view/10.1093/ref: odnb/9780198614128.001.0001/odnb-9780198614128-e-95963.

Crowe, M. J., "Sir John Frederick William Herschel (1792–1871)," in *Oxford Dictionary of National Biography*, Oxford, Oxford University Press, 2009. www.oxforddnb.com/view/10.1093/ref:odnb/9780198614128.001.0001/ odnb-9780198614128-e-13101.

Enros, P. C., "The Analytical Society (1812–1813): Precursor of the Renewal of Cambridge Mathematics," *Historia Mathematica*, 10 (1983), 24–47.

Fisch, M., *Creativity Undecided: Towards a History and Philosophy of Scientific Agency*, Chicago, University of Chicago Press, 2017.

Verburgt, L. M., "Duncan F. Gregory, William Walton and the Development of British Algebra: 'Algebraical Geometry,' 'Geometrical Algebra,' Abstraction," *Annals of Science*, 73 (2016), 40–67.

Chapter 3: John Herschel's Astronomy

Ashworth, W., "The Calculating Eye: Baily, Herschel, and the Business of Astronomy," *British Journal for the History of Science*, 27 (1994), 409–41.

Case, S., *Making Stars Physical: The Astronomy of Sir John Herschel*, Pittsburgh, PA, University of Pittsburgh Press, 2018.

Case, S., "A 'Confounded Scrape': John Herschel, Neptune, and Naming the Satellites of the Outer Solar System," *Journal for the History of Astronomy*, 50 (2019), 306–25.

Crowe, M. J., *The Extraterrestrial Life Debate, 1750–1900*, Mineola, NY, Dover, 1999.

Herschel, J. F. W., *Outlines of Astronomy*, London, Longman, Brown, Green, and Longmans, 1849.

Hoskin, M., "John Herschel's Cosmology," *Journal for the History of Astronomy*, 18 (1987), 1–34.

Chapter 4: Stargazer at World's End: John Herschel at the Cape, 1833–1838

Evans, D. et al. (eds.), *Herschel at the Cape: Diaries and Correspondence of Sir John Herschel, 1834–1838*, Austin, University of Texas Press, 1969.

Musselman, E. G., "Swords into Ploughshares: John Herschel's Progressive View of Astronomical and Imperial Governance," *British Journal for the History of Science*, 31 (1998), 419–36.

Ruskin, S., *John Herschel's Cape Voyage: Private Science, Public Imagination, and the Ambitions of Empire*, Burlington, VT, Ashgate, 2004.

Warner, B., "Sir John Herschel at the Cape of Good Hope," in B. Warner (ed.), *John Herschel 1792–1992: Bicentennial Symposium*, Cape Town, Royal Society of South Africa, 1994, 19–55.

Warner, B., and J. Rourke, *Flora Herscheliana: Sir John and Lady Herschel at the Cape 1834 to 1838*, Johannesburg, The Brenthurst Press, 1998.

Chapter 5: Herschel's Philosophy of Science: The *Preliminary Discourse on the Study of Natural Philosophy* and Beyond

Ashworth, W. J., "The Calculating Eye: Baily, Herschel, Babbage and the Business of Astronomy," *British Journal for the History of Science*, 27 (1994), 409–41.

Bolt, M. P., "John Herschel's Natural Philosophy: On the Knowing of Nature and the Nature of Knowing in Early Nineteenth-Century Britain," PhD dissertation, University of Notre Dame, 1998.

Cobb, A. D., "Is John F. W. Herschel an Inductivist about Hypothetical Inquiry?" *Perspectives on Science*, 20 (2012), 409–39.

Hankins, T. L., "A 'Large and Graceful Sinuosity': John Herschel's Graphical Method," *Isis*, 97 (2006), 605–33.

Herschel, J. F. W., "Quetelet on Probabilities," *Edinburgh Review*, 92 (1850), 1–57.

Secord, J. A., "The Conduct of Everyday Life: John Herschel's *Preliminary Discourse on the Study of Natural Philosophy*," in *Visions of Science: Books and Readers at the Dawn of the Victorian Age*, Chicago, University of Chicago Press, 2014, 80–106.

Yeo, R., "Scientific Method and the Rhetoric of Science in Britain, 1830–1917," in J. Schuster and R. Yeo (eds.), *The Politics and Rhetoric of Scientific Method: Historical Studies*, Dordrecht, Reidel, 1986, 259–97.

Yeo, R., "Reviewing Herschel's *Discourse*," *Studies in History and Philosophy of Science Part A*, 20 (1989), 541–52.

Chapter 6: Drawing Observations Together: John Herschel and the Art of Drawing in Scientific Observations

Fiorentini, E., "Practices of Refined Observation: The Conciliation of Experience and Judgement in John Herschel's Discourse and in His Drawings," in E. Fiorentini (ed.), *Observing Nature – Representing Experience: The Osmotic Dynamics of Romanticism 1800–1850*, Berlin, Dietrich Reimer Verlag, 2007, 19–42.

Schaaf, L. J., *Tracings of Light: Sir John Herschel & the Camera Lucida*, San Francisco, Friends of Photography, 1989.

Warner, B., *Cape Landscapes: Sir John Herschel's Sketches 1834–1838*, Cape Town, University of Cape Town Press, 2006.

Warner, B., and J. Rourke, *Flora Herscheliana: Sir John and Lady Herschel at the Cape 1834 to 1838*, Johannesburg, The Brenthurst Press, 1998.

Chapter 7: Photology, Photography, and Actinochemistry: The Photographic Work of John Herschel

Hentschel, K., *Mapping the Spectrum: Techniques of Visual Representation in Research and Teaching*, Oxford, Oxford University Press, 2002.

Schaaf, L. J., *Out of the Shadows: Herschel, Talbot and the Invention of Photography*, New Haven, CT, Yale University Press, 1992.

Ware, M., *Cyanotype: History, Science and Art of Photographic Printing in Prussian Blue*, London, Science Museum, 1999.

Wilder, K., "A Note on the Science of Photography: Reconsidering the Invention Story," in T. Sheehan and A. Zervigon (eds.), *Photography and Its Origins*, New York, Routledge, 2015, 208–21.

Chapter 8: Herschel's Planet: Earth in Cosmic Perspective

Good, G. A., "John Herschel's Geology: The Cape of Good Hope in the 1830s," in J. Buchwald and L. Stewart (eds.), *The Romance of Science: Essays in Honour of Trevor H. Levere*, Cham, Springer, 2017, 135–50.

Good, G. A., *On the Trail of Charles Darwin and John Herschel: The Cape in the 1830s*, 35th International Geological Congress Field Trip Guide, 2016.

Chapter 9: John Herschel and Standardization

Case, S., "'Land-marks of the Universe': John Herschel against the Background of Positional Astronomy," *Annals of Science*, 72, 2015.

Cawood, J., "The Magnetic Crusade: Science and Politics in Early Victorian Britain," *Isis*, 70 (1979), 493–518.

Gillin, E., and F. Gribenski, "The Politics of Musical Standardization in Nineteenth-Century France and Britain," *Past and Present*, 251 (2021), 153–87.

Herschel, J. F. W. (ed.), *Manual of Scientific Enquiry; Prepared for the Use of Her Majesty's Navy: and Adapted for Travellers in General*, London, John Murray, 1849.

Herschel, J. F. W., *On Musical Scales*, London, W. Clowes and Sons, 1868.

Herschel, J. F. W., *Familiar Lectures on Scientific Subjects*, New York, George Routledge & Sons, 1869.

Schaffer, S., "Metrology, Metrication, and Victorian Values," in B. Lightman (ed.), *Victorian Science in Context*, Chicago, University of Chicago Press, 1997, 438–74.

Chapter 10: John Herschel and Politics

Ashworth, W. J., "The Calculating Eye: Baily, Herschel, Babbage and the Business of Astronomy," *British Journal for the History of Science*, 27 (1994), 409–41.

Buttmann, G., *The Shadow of the Telescope: A Biography of John Herschel*, London, Lutterworth Press, 1974.

Crowe, M. J., D. R. Dyck, and J. R. Kevin (eds.), *A Calendar of the Correspondence of Sir John Herschel*, Cambridge, Cambridge University Press, 1998.

Schweber, S. S. (ed.), *Aspects of the Life and Thought of John Herschel*, New York, Arno Press, 1981.

Chapter 11: John Herschel's Methodology in the Scientific Community

Bellon, R., "Sacrifice in Service to Truth: The Epistemic Virtues of Victorian British Science," in E. Ratti and T. A. Stapleford (eds.), *Science, Technology, and Virtues*, Oxford, Oxford University Press, 2021, 17–36.

Cannon, W. F., "The Impact of Uniformitarianism: Two Letters from John Herschel to Charles Lyell, 1836–1837," *Proceedings of the American Philosophical Society*, 105 (1961), 307–8.

Gooding, D., "'He Who Proves, Discovers': John Herschel, William Pepys and the Faraday Effect," *Notes and Records of the Royal Society of London*, 39 (1985), 229–44.

Ross, S., "John Herschel on Faraday and on Science," *Notes and Records of the Royal Society of London*, 33 (1978), 77–82.

Secord, J. A., "The Conduct of Everyday Life: John Herschel's *Preliminary Discourse on the Study of Natural Philosophy*," in *Visions of Science: Books and Readers at the Dawn of the Victorian Age*, Chicago, University of Chicago Press, 2014, 80–106.

Index